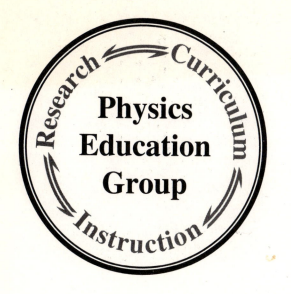

Tutorials in Introductory Physics

First Edition

D0218076

Lillian C. McDermott, Peter S. Shaffer
and the Physics Education Group

Department of Physics
University of Washington

Prentice Hall
Upper Saddle River, New Jersey 07458

Prentice Hall Series in Educational Innovation

EDITOR IN CHIEF: *John Challice*
ACQUISITIONS EDITOR: *Alison Reeves*
EXECUTIVE MANAGING EDITOR: *Kathleen Schiaparelli*
ASSISTANT MANAGING EDITOR: *Beth Sturla*
PRODUCTION EDITORS: *Shari Toron and Susan Fisher*
ART DIRECTOR: *Jayne Conte*
COVER DESIGNER: *Bruce Kenselaar*
MANUFACTURING MANAGER: *Trudy Pisciotti*
ASSISTANT MANUFACTURING MANAGER: *Michael Bell*
VICE PRESIDENT OF PRODUCTION AND MANUFACTURING: *David W. Riccardi*

© 2002 by Prentice-Hall, Inc.
Upper Saddle River, New Jersey 07458

Printed in the United States of America
10 9 8 7 6 5 4 3 2 1

ISBN:0-13-065364-0

Pearson Education Ltd., *London*
Pearson Education Australia Pty. Limited, *Sydney*
Pearson Education Singapore, Pte. Ltd.
Pearson Education North Asia Ltd., *Hong Kong*
Pearson Education Canada Ltd., *Toronto*
Pearson Educación de Mexico, S.A. de C.V.
Pearson Education—Japan, *Tokyo*
Pearson Education Malaysia, Pte. Ltd.

Preface

Tutorials in Introductory Physics is a set of instructional materials intended to supplement the lectures and textbook of a standard introductory physics course. The emphasis in the tutorials is on the development of important physical concepts and scientific reasoning skills, not on solving the standard quantitative problems found in traditional textbooks.

There is increasing evidence that after instruction in a typical course, many students are unable to apply the physics formalism that they have studied to situations that they have not expressly memorized. In order for meaningful learning to occur, students need more assistance than they can obtain through listening to lectures, reading the textbook, and solving standard quantitative problems. It can be difficult for students who are studying physics for the first time to recognize what they do and do not understand and to learn to ask themselves the types of questions necessary to come to a functional understanding of the material. *Tutorials in Introductory Physics* provides a structure that promotes the active mental engagement of students in the process of learning physics. Questions in the tutorials guide students through the reasoning necessary to construct concepts and to apply them in real-world situations. The tutorials also provide practice in interpreting various representations (*e.g.,* verbal descriptions, diagrams, graphs, and formulas) and in translating back and forth between them. For the most part, the tutorials are intended to be used after concepts have been introduced in the lectures and the laboratory, although most can serve to introduce the topic as well.

The tutorials comprise an integrated system of pretests, worksheets, homework assignments, and post-tests. The tutorial sequence begins with a pretest. These are usually on material already presented in lecture or textbook but not yet covered in tutorial. The pretests help students identify what they do and not understand about the material and what they are expected to learn in the upcoming tutorial. They also inform the instructors about the level of student understanding. The worksheets, which consist of carefully sequenced tasks and questions, provide the structure for the tutorial sessions. Students work together in small groups, constructing answers for themselves through discussions with one another and with the tutorial instructors. The tutorial instructors do not lecture but ask questions designed to help students find their own answers. The tutorial homework reinforces and extends what is covered in the worksheets. For the tutorials to be most effective, it is important that course examinations include questions that emphasize the concepts and reasoning skills developed in the tutorials.

The tutorials are primarily designed for a small class setting but have proved to be adaptable to other instructional environments. The curriculum has been shown to be effective for students in regular and honors sections of calculus-based and algebra-based physics.

The tutorials have been developed through an iterative cycle of: research on the learning and teaching of physics, design of curriculum based on this research, and assessment through rigorous pretesting and post-testing in the classroom. *Tutorials in Introductory Physics* has been developed and tested at the University of Washington and pilot-tested at other colleges and universities.

Comments on the First Edition

Ongoing research has led to modifications to the tutorials and associated homework in the Preliminary Edition of *Tutorials in Introductory Physics*. The First Edition incorporates these changes and also includes several new tutorials on topics covered in the Preliminary Edition. In addition, the First Edition contains a new section with tutorials on topics in hydrostatics, thermal physics, and modern physics.

Acknowledgments

Tutorials in Introductory Physics is the product of close collaboration by many members of the Physics Education Group at the University of Washington. In particular, Paula Heron and Stamatis Vokos, faculty in physics, have played an important role in the development of many tutorials. Significant contributions to the First Edition have also been made by current and former graduate students and post-doctoral research associates: Bradley Ambrose, Andrew Boudreaux, Matt Cochran, Gregory Francis, Stephen Kanim, Christian Kautz, Michael Loverude, Luanna G. Ortiz, Mel Sabella, Rachel Scherr, Mark Somers, John Thompson, and Karen Wosilait. Others, whose work on the Preliminary Edition enriched the First Edition include: Chris Border, Patricia Chastain, Randal Harrington, Pamela Kraus, Graham Oberem, Daryl Pedigo, Tara O'Brien Pride, Christopher Richardson, and Richard Steinberg. Lezlie S. DeWater and Donna Messina, experienced K-12 teachers, have provided many useful insights and suggestions. The assistance of Joan Valles in coordinating the work of the Physics Education Group is deeply appreciated.

The collaboration of other colleagues in the Physics Department has been invaluable. Faculty in the introductory calculus-based sequence, and graduate and undergraduate students who have served as tutorial instructors have made many useful comments. Contributions have also been made by many long-term and short-term visitors to our group. Physics instructors who have pilot-tested the tutorials and have provided valuable feedback over an extended period of time include: John Christopher (University of Kentucky), Romana Crnkovic (Minot State University), William Duxler (Los Angeles Pierce College), Robert Endorf (University of Cincinnati), Gregory Francis (Montana State University), James Freericks and Amy Liu (Georgetown University), Gary Gladding (University of Illinois, Urbana-Champaign), Gregory Kilcup (The Ohio State University), Heidi Mauk (The United States Air Force Academy), Eric Mazur (Harvard University), James Poth (Miami University), and E.F. Redish (University of Maryland).

We thank our editor, Alison Reeves, for her encouragement and advice. We also gratefully acknowledge the support of the National Science Foundation, which has enabled the Physics Education Group to conduct the ongoing, comprehensive program of research, curriculum development, and instruction that has produced *Tutorials in Introductory Physics*. The tutorials have also benefited from the concurrent development of *Physics by Inquiry* (©1996 John Wiley & Sons, Inc.). *Tutorials in Introductory Physics* and *Physics by Inquiry* share a common research base and portions of each have been adapted for the other.

Table of Contents

Part I: Mechanics

Kinematics

Newton's laws

Energy and momentum

Rotation

Part II: Electricity and magnetism

Electrostatics

Electric circuits

Magnetism

Electromagnetism

Part III: Waves

Part IV: Optics

Geometrical optics

Physical optics

Part V: Selected topics

Hydrostatics

Thermodynamics

Modern Physics

Mechanics

I. Motion with constant speed

Each person in your group should obtain a ruler and at least one ticker tape segment from the staff. All the tape segments were generated using the same ticker timer. Do not write on or fold the tapes. If a ticker timer is available, examine it so that you are familiar with how it works.

A. Describe the motion represented by your segment of tape. Explain your reasoning.

B. Compare your tape segment with those of your partners.

How does the time taken to generate one of the short tape segments compare to the time taken to generate one of the long tape segments? Explain your answer.

Describe how you could use your answer above to arrange the tape segments in order by speed.

C. Suppose the ticker timer that made the dots strikes the tape every $1/60^{th}$ of a second.

How far did the object that generated your tape segment move in: $1/60^{th}$ of a second? $2/60^{th}$ of a second? $3/60^{th}$ of a second? Explain your answer.

Predict how far the object would move in: 1 second, $1/120^{th}$ of a second. Explain the assumption(s) you used to make your predictions.

D. In your own words, describe a procedure you could use to calculate the speed of an object.

E. Determine the speed of the object that generated each of your tapes. Record your answers below.

Give an interpretation of the speed of the object, *i.e.,* explain the meaning of the number you just calculated. Do not use the word "speed" in your answer. (*Hint:* Which of the distances that you calculated in part C is numerically equal to the speed?)

Write the speed of the object that generated each tape on a small piece of paper and attach the paper to the tape. Express your answer in terms of centimeters and seconds.

Tutorials in Introductory Physics
McDermott, Shaffer, & P.E.G., U. Wash.

©Prentice Hall, Inc.
First Edition, 2002

F. A motion that generates a sequence of evenly-spaced dots on a ticker tape is called motion with *constant speed*. Explain the assumption about the motion that is being made when this phrase is applied.

Discuss with your partners whether the object that generated your tape was moving with constant velocity.

G. A model train moving with constant velocity travels 60 cm for every 1.5 s that elapses. Answer the questions below and discuss your reasoning with your partners.

1. Is there a name that is commonly given to the quantity represented by the number 40? ($40 = 60/1.5$) If so, what is the name?

To denote the quantity completely, what additional information must be given besides the number 40?

How would you *interpret* the number 40 in this instance? (*Note:* A name is *not* an interpretation. Your response should be in terms of centimeters and seconds.)

Use your interpretation (not algebra) to find the distance the train moves in 2.5 s.

2. Is there a name that is commonly given to the quantity represented by the number 0.025? ($0.025 = 1.5/60$) If so, what is the name?

How would you *interpret* the number 0.025?

Use your interpretation (not algebra) to find the time it takes the train to move 90 cm.

II. Motion with varying speed

A. In the space below, sketch a possible ticker tape resulting from motion with varying speed and write a description of the motion.

How can you tell from your diagram that the motion has varying speed?

B. Together with your classmates, take your ticker tape segments and arrange yourselves in a line, ranked according to the speed of the segments. Discuss the following questions as a class.

Compare your segment of ticker tape to neighboring tape segments. What do you observe?

Compare the smallest and largest speeds. What do you observe?

C. Based on your observations of your tape segment and the tape segments of other members of your class, answer the following questions.

Is each small tape segment a part of a motion with constant or varying speed?

Did your examination of a single, small tape segment reveal whether the entire motion that generated the tape had constant or varying speed?

D. Review your earlier interpretation of the speed for your small tape segment. (See section I.) Is that interpretation valid for the entire motion that generated the tape?

Based on the speed for your piece of tape, could you successfully predict how far the object would move in: $1/60^{th}$ s? $2/60^{th}$ s? 1 s?

How can you modify the interpretation of the speed so that it applies even to motion with varying speed?

What name is given to a speed that is interpreted in this way?

E. Suppose you selected two widely separated dots on the ticker tape assembled in part B. What would you call the number you would obtain if you divided the distance between the dots by the time it took the object to move between the dots?

How would you interpret this number?

In this tutorial, you will use a motion detector to graph your motion and to investigate how motion can be described in terms of position, velocity, and acceleration. See your instructor for instructions on using the equipment.

General tips

When using a motion detector:

• Stay in line with the detector and do not swing your arms. For best results, take off bulky sweaters or other loose-fitting clothing. You may find it helpful to hold a large board in front of you in order to present a larger target for the detector.

• Do not stand closer than about 0.5 meter or farther than 4.0 meters from the detector.

• It is difficult to obtain good a versus t graphs with the motion detector. Discuss any questions about your a versus t graphs with an instructor.

Instructions

In each of the following problems, you will be given one of the following descriptions of a motion:

• a written description, or

• an x versus t, v versus t, or a versus t graph.

Predict the other three descriptions of the motion, then use the motion detector to check your answers. Check your predictions one-by-one, instead of checking several problems at once. In addition, answer the questions posed at the bottom of each page.

Note: So that your graphs emphasize important features, draw them in an idealized form rather than showing many small wiggles.

Example: The problem below has been worked as an example. Use the motion detector to verify the answers.

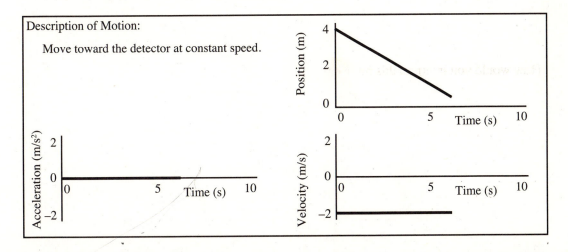

A. The computer program assumes a particular coordinate system. Describe this coordinate system.

Tutorials in Introductory Physics
McDermott, Shaffer, & P.E.G., U. Wash.

©Prentice Hall, Inc.
First Edition, 2002

B.

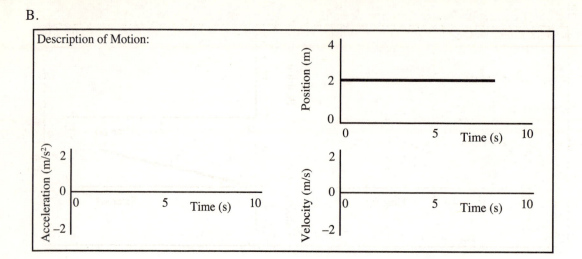

Description of Motion:

C.

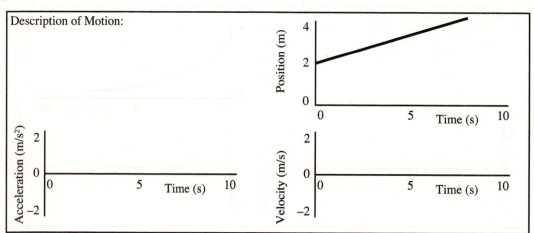

Description of Motion:

D.

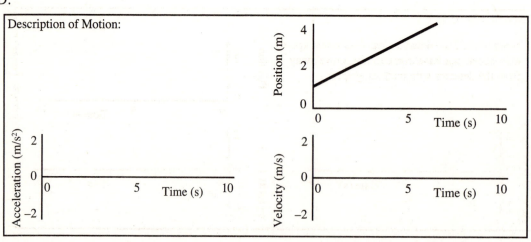

Description of Motion:

E. How are the motions in parts C and D similar? How do they differ? How are the graphs similar? How do they differ?

Tutorials in Introductory Physics
McDermott, Shaffer, & P.E.G., U. Wash.

©Prentice Hall, Inc.
First Edition, 2002

F.

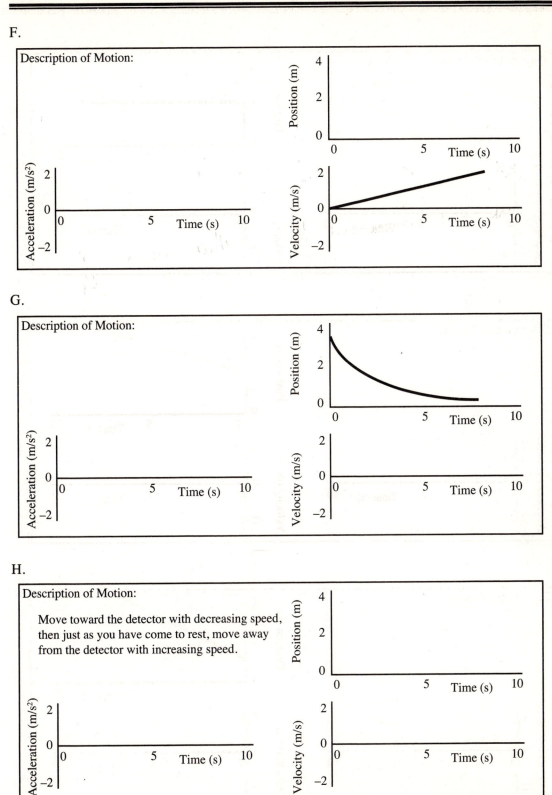

G.

H.

I. How do the acceleration graphs for F, G, and H compare? Is it possible to have: a positive acceleration and slow down? a negative acceleration and speed up?

J.

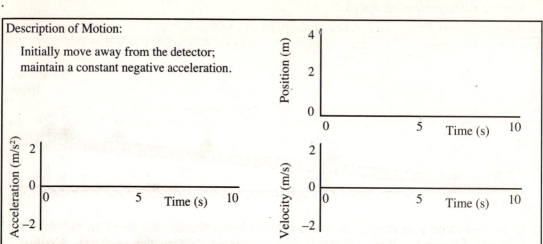

Description of Motion:

Initially move away from the detector; maintain a constant negative acceleration.

K.

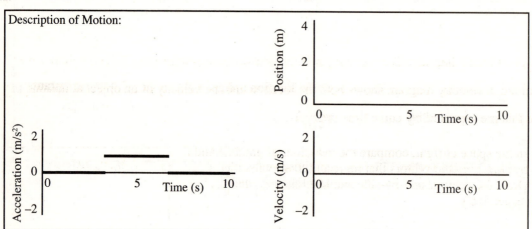

Description of Motion:

L.

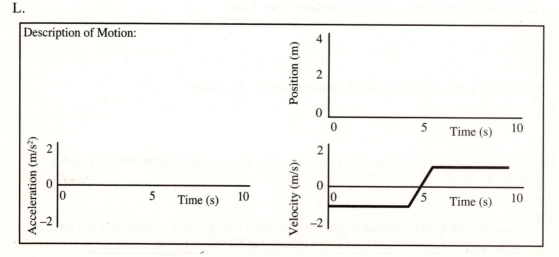

Description of Motion:

M. The term *decelerate* is often used to indicate that an object is slowing down. Does this term indicate the sign of the acceleration?

Tutorials in Introductory Physics
McDermott, Shaffer, & P.E.G., U. Wash.

©Prentice Hall, Inc.
First Edition, 2002

I. Motion with decreasing speed

The diagram below represents a strobe photograph of a ball as it rolls *up* a track. (In a strobe photograph, the position of an object is shown at instants separated by *equal time intervals*.)

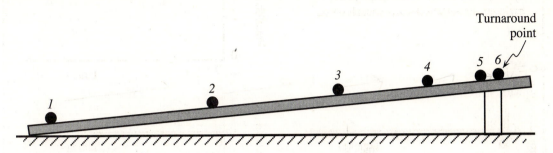

A. Draw vectors on your diagram that represent the instantaneous velocity of the ball at each of the labeled locations. If the velocity is zero at any point, indicate that explicitly. Explain why you drew the vectors as you did.

We will call diagrams like the one you drew above *velocity diagrams*. Unless otherwise specified, a velocity diagram shows both the location and the velocity of an object at instants in time that are separated by equal time intervals.

B. In the space at right, compare the velocities at points *1* and *2* by sketching the vectors that represent those velocities. Draw the vectors side-by-side and label them $\vec{v}_1$ and $\vec{v}_2$, respectively.

$\vec{v}_1$, $\vec{v}_2$, and $\Delta\vec{v}$

Draw the vector that must be *added* to the velocity at the earlier time to equal the velocity at the later time. Label this vector $\Delta\vec{v}$.

Why is the name *change in velocity* appropriate for this vector?

How does the direction of the change in velocity vector compare to the direction of the velocity vectors?

Would your answer change if you were to select two *different* consecutive points (*e.g.,* points *3* and *4)* while the ball was slowing down? Explain.

Tutorials in Introductory Physics
McDermott, Shaffer, & P.E.G., U. Wash.

©Prentice Hall, Inc.
First Edition, 2002

How would the magnitude of the change in velocity vector between points *1* and *2* compare to the magnitude of the change in velocity vector between two *different* consecutive points (*e.g.*, points *3* and *4*)? Explain. (You may find it useful to refer to the graph of velocity versus time for the motion.)

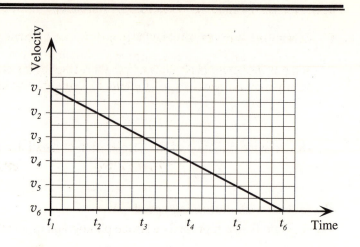

Note: The positive direction has been chosen to be *up* the track.

C. Consider the change in velocity vector between two points on the velocity diagram that are not consecutive, *e.g.*, points *1* and *4*.

Is the direction of the change in velocity vector different than it was for consecutive points? Explain.

Is the length of the change in velocity vector different than it was for consecutive points? If so, how many times larger or smaller is it than the corresponding vector for consecutive points? Explain.

D. Use the definition of acceleration to draw a vector in the space at right that represents the acceleration of the ball between points *1* and *2*.

Acceleration vector

How is the direction of the acceleration vector related to the direction of the change in velocity vector? Explain.

E. Does the acceleration change as the ball rolls up the track?
Would the acceleration vector you obtain differ if you were to choose (1) two different successive points on your diagram or (2) two points that are not consecutive? Explain.

F. Generalize your results thus far to answer the following question:

What is the relationship between the direction of the acceleration and the direction of the velocity for an object that is moving in a straight line and slowing down? Explain.

Describe the direction of the acceleration of a ball that is rolling up a straight incline.

II. Motion with increasing speed

The diagram below represents a strobe photograph of a ball as it rolls *down* the track.

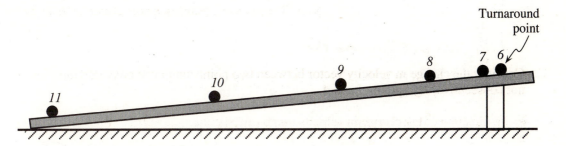

A. Choose two successive points. In the space at right, sketch the velocity vectors corresponding to those points. Draw the vectors side-by-side and label them $\vec{v}_i$ and $\vec{v}_f$, respectively.

Determine the vector that must be added to the velocity at the earlier time to equal the velocity at the later time. Is the name *change in velocity* appropriate for this vector?

How does the direction of the change in velocity vector compare to the direction of the velocity vectors in this case?

$\vec{v}_i$, $\vec{v}_f$, and $\Delta \vec{v}$

Would your answer change if you were to select two *different* points during the time that the ball was speeding up? Explain.

B. In the space at right, draw a vector to represent the acceleration of the ball between the points chosen above.

How is the direction of the change in velocity vector related to the direction of the acceleration vector? Explain.

Acceleration vector

©Prentice Hall, Inc.
First Edition, 2002

Generalize your results thus far to answer the following question:

What is the relationship between the direction of the acceleration and the direction of the velocity for an object that is moving in a straight line and speeding up? Explain.

Describe the direction of the acceleration of a ball that is rolling down a straight incline.

III. Motion that includes a change in direction

Complete the velocity diagram below for the portion of the motion that includes the turnaround.

A. Choose a point *before* the turnaround and another *after*.

In the space below, draw the velocity vectors and label them $\vec{v}_i$ and $\vec{v}_f$.

Draw the vector that must be added to the velocity at the earlier time to obtain the velocity at the later time.

Is the name *change in velocity* that you used in sections I and II also appropriate for this vector?

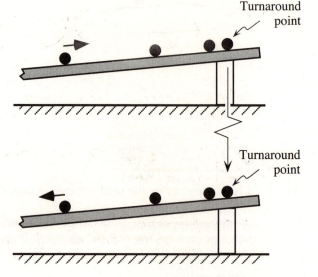

B. Suppose that you had chosen the turnaround as one of your points.

What is the velocity at the turnaround point?

Would this choice affect the direction of the change in velocity vector? Explain why or why not.

$\vec{v}_i$, $\vec{v}_f$, and $\Delta\vec{v}$

C. In the space at right, draw a vector that represents the acceleration of the ball between the points you chose in part B above.

Compare the direction of the acceleration of the ball at the turnaround point to that of the ball as it rolls: (1) up the track and (2) down the track.

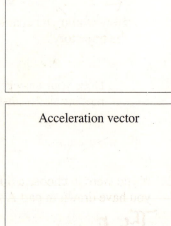

Acceleration vector

Tutorials in Introductory Physics
McDermott, Shaffer, & P.E.G., U. Wash.

©Prentice Hall, Inc.
First Edition, 2002

I. Velocity

An object is moving around an oval track. Sketch the trajectory of the object on a large sheet of paper. (Make your diagram *large*.)

A. Choose a point to serve as an origin for your coordinate system. Label that point O (for origin). Select two locations of the object that are about one-eighth of the oval apart and label them A and B.

1. Draw the position vectors for each of the two locations A and B and draw the vector that represents the displacement from A to B.

Copy your group's drawing in this space after discussion.

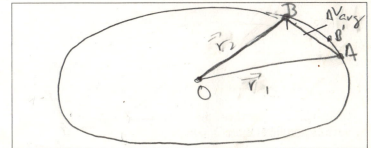

2. Describe how to use the displacement vector to determine the direction of the average velocity of the object between A and B. Draw a vector to represent the average velocity.

The displacement vector and the average velocity of the object are both calculated using the change in position divided by the change in time.

3. Choose a point on the oval between points A and B, and label that point B'.

As point B' is chosen to lie closer and closer to point A, does the direction of the average velocity over the interval AB' change? If so, how?

As B' gets closer to A the velocity vector shifts toward.

4. Describe the direction of the instantaneous velocity of the object at point A.

The instantaneous velocity of an object at point A is tangent to the oval's point.

How would you characterize the direction of the instantaneous velocity at *any* point on the trajectory? The direction of the instantaneous velocity at any point on the trajectory is tangent to the oval's point.

Does your answer depend on whether the object is speeding up, slowing down, or moving with constant speed? Explain.

No, because instantaneous velocity is always tangent to the position the oval

B. If you were to choose a different origin for the coordinate system, which of the vectors that you have drawn in part A would change and which would not change?

The position vectors, point B', and origin would change, and the average velocity would stay the same.

Tutorials in Introductory Physics
McDermott, Shaffer, & P.E.G., U. Wash.

©Prentice Hall, Inc.
First Edition, 2002

II. Acceleration for motion with constant speed

Suppose that the object in section I is moving around the track at *constant speed*. Draw vectors to represent the velocity at two points on the track that are relatively close together. (Draw your vectors *large*.) Label the two points *C* and *D*.

A. On a *separate* part of your paper, copy the velocity vectors $\vec{v}_C$ and $\vec{v}_D$. From these vectors, determine the *change in velocity vector*, $\Delta\vec{v}$.

1. Is the angle formed by the "head" of $\vec{v}_C$ and the "tail" of $\Delta\vec{v}$ *greater than, less than*, or *equal to* 90°?

    The angle is less then 90°

    As point *D* is chosen to lie closer and closer to point *C*, does the above angle *increase, decrease*, or *remain the same?* Explain how you can tell.

    The angle above decreases, because the triangle gets smaller.

    Does the above angle approach a *limiting* value? If so, what is its limiting value?

    Yes, the limiting value is 0.

2. Describe how to use the change in velocity vector to determine the average acceleration of the object between *C* and *D*. Draw a vector to represent the average acceleration between points *C* and *D*.

    The average acceleration is the rate of change of the object's velocity, then averaged.

    What happens to the magnitude of $\Delta\vec{v}$ as point *D* is chosen to lie closer and closer to point *C?* Does the acceleration change in the same way? Explain.

    The magnitude decreases, the acceleration does change in the same way, because if the magnitude and acceleration have a proportional relationship.

    Consider the direction of the acceleration at point *C*. Is the angle between the acceleration vector and the velocity vector *greater than, less than*, or *equal to* 90°? (*Note:* Conventionally, the angle between two vectors is defined as the angle formed when they are placed "tail-to-tail.")

    The vector is 90°.

B. Suppose you were to choose a new point on the trajectory where the *curvature is different* from that at point *C*.

Is the magnitude of the acceleration at the new point *greater than, less than,* or *equal to* the magnitude of the acceleration at point *C?* Explain.

The magnitude would be the same as the magnitude of the curvature at point C, because the magnitude won't change.

Describe the direction of the acceleration at the new point.

The direction of acceleration is now points toward the center of the circle.

⇨ Check your reasoning for section II with a tutorial instructor before proceeding.

III. Acceleration for motion with changing speed

Suppose that the object is *speeding up* as it moves around the oval track. Draw vectors to represent the velocity at two points on the track that are relatively close together. (Draw your vectors *large*.) Label the two points *E* and *F*.

A. On a *separate* part of your paper, copy the velocity vectors $\vec{v}_E$ and $\vec{v}_F$. From these vectors, determine the change in velocity vector, $\Delta\vec{v}$.

1. Is the angle θ, formed by the head of $\vec{v}_E$ and the tail of $\Delta\vec{v}$, *greater than, less than,* or *equal to* 90°? *The angle is less than 90°.*

Consider how θ changes as point *F* is chosen to lie closer and closer to point *E*.

What value or range of values is possible for this angle for an object that is speeding up? Explain.

What happens to the magnitude of $\Delta\vec{v}$ as point *F* is chosen to lie closer and closer to point *E*? *The magnitude decreases.*

2. Describe how you would determine the acceleration of the object at point *E*.

The acceleration of the object at point E is defined by $a \equiv \frac{dv}{dt}$

Consider the direction of the acceleration at point *E*. Is the angle between the acceleration vector and the velocity vector (placed "tail-to-tail") *greater than, less than,* or *equal to* 90°?

The vector is greater than 90°.

B. Suppose the object started *from rest* at point *E* and moved towards point *F* with increasing speed. How would you find the acceleration at point *E*?

$$\frac{\vec{v}_f - \vec{v}_i}{t_f - t_i} = \bar{a}$$

Describe the direction of the acceleration of the object at point *E*.

The acceleration is parallel to the line going E to F

C. At several points on each of the diagrams below, draw a vector that represents the acceleration of the object.

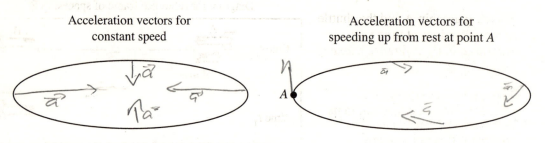

Acceleration vectors for constant speed	Acceleration vectors for speeding up from rest at point *A*
Top view diagram	Top view diagram

Characterize the direction of the acceleration at each point on the trajectory for each case.

The direction of the acceleration on the constant speed oval is toward the midpoint. The direction for the acceleration for the speeding up is toward no point.

Is the acceleration directed toward the "center" of the oval at every point on the trajectory for either of these cases?

No the acceleration is not directed toward the center of either of the ovals.

Sketch arrows to show the direction of the acceleration for the following trajectories:

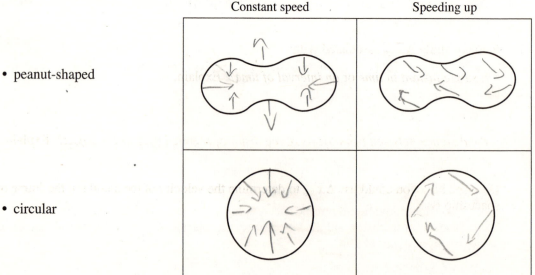

	Constant speed	Speeding up
• peanut-shaped		
• circular		

Tutorials in Introductory Physics
McDermott, Shaffer, & P.E.G., U. Wash.

©Prentice Hall, Inc.
First Edition, 2002

I. Position and displacement relative to different observers

Two spaceships, A and B, move toward one another as shown. At time t_i, spaceship A launches a shuttle craft toward spaceship B. At time t_f, the shuttle reaches spaceship B.

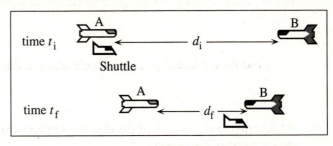

A. The second diagram at right shows the positions of spaceships A, B, and the shuttle at time t_i in the reference frame of spaceship B.

   Sketch spaceship A and the shuttle at their positions at time t_f *as measured in the reference frame of spaceship B*.

   Explain how the diagram is consistent with the fact that in its own frame of reference, spaceship B is not moving.

Diagram for reference frame of spaceship B

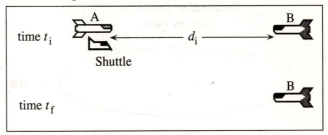

In the box at right, draw and label vectors for the following quantities:

$\vec{x}_S^{(i)}$, $\vec{x}_S^{(f)}$, and $\Delta\vec{x}_S$ in frame of spaceship B

- $\vec{x}_{S,B}^{(i)}$ (the *initial position* of the shuttle in the frame of spaceship B)

- $\vec{x}_{S,B}^{(f)}$ (the *final position* of the shuttle in the frame of spaceship B)

- $\Delta\vec{x}_{S,B}$ (the *displacement*, or *change in position*, of the shuttle in the frame of spaceship B).

Is the quantity $\Delta\vec{x}_{S,B}$ associated with:

- *a single instant in time* or *an interval of time?* Explain.

- *the distance between two objects* or *the distance traveled by a single object?* Explain.

Describe how you could use $\Delta\vec{x}_{S,B}$ to determine the velocity of the shuttle in the frame of spaceship B.

B. The picture of the spaceships and shuttle from the previous page is reproduced at right.

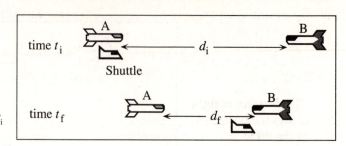

The diagram below the picture shows the positions of the two spaceships and the shuttle at time t_i in the reference frame of spaceship A.

Sketch spaceship A, B, and the shuttle at their positions at time t_f *as measured in the reference frame of spaceship A.*

Diagram for reference frame of spaceship A

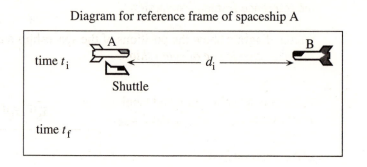

In the box at right, draw and label vectors for $\vec{x}_{S,A}^{(i)}$, $\vec{x}_{S,A}^{(f)}$, and $\Delta \vec{x}_{S,A}$.

$\vec{x}_S^{(i)}$, $\vec{x}_S^{(f)}$, and $\Delta \vec{x}_S$ in frame of spaceship A

Describe how you could use $\Delta \vec{x}_{S,A}$ to determine the velocity of the shuttle in the frame of spaceship A.

Is the magnitude of the velocity of the shuttle in the frame of spaceship A *greater than, less than,* or *equal to* the magnitude of the velocity of the shuttle in the frame of spaceship B? Explain.

C. Rank d_i, d_f, $|\Delta \vec{x}_{S,A}|$, and $|\Delta \vec{x}_{S,B}|$ in order of decreasing magnitude. Make sure your ranking is consistent with your previous results.

Tutorials in Introductory Physics
McDermott, Shaffer, & P.E.G., U. Wash.

©Prentice Hall, Inc.
First Edition, 2002

D. Spaceship C moves so as to remain a fixed distance behind spaceship B at all times.

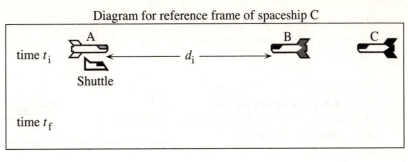

Diagram for reference frame of spaceship C

The diagram at right shows the positions of the three spaceships and the shuttle at time t_i in the reference frame of spaceship C.

On the diagram show the positions of the spaceships and the shuttle at time t_f *as measured in the reference frame of spaceship C.*

In the box at right, draw and label vectors for $\vec{x}_{s,C}^{(i)}$, $\vec{x}_{s,C}^{(f)}$, and $\Delta \vec{x}_{s,C}$.

$\vec{x}_S^{(i)}, \vec{x}_S^{(f)},$ and $\Delta \vec{x}_S$ in frame of spaceship C

Is the magnitude of the displacement of the shuttle in the frame of spaceship C *greater than, less than,* or *equal to* the magnitude of the displacement of the shuttle in the frame of spaceship B? Explain.

E. Consider the following statement:

"The displacement of the shuttle is greater relative to spaceship C than it is relative to spaceship B. At time t_f, the shuttle is right next to spaceship B, but it is still a large distance away from spaceship C."

Do you agree or disagree? Explain.

If all displacements of an object (such as the shuttle) are measured to have the same value by two different observers, those observers are said to be *in the same frame of reference*.

F. State which of the spaceships, if any, are in the same frame of reference. Explain.

Generalize your answer to describe the conditions under which two observers are in the same frame of reference.

II. Relative Velocity

A car and a truck move on a straight road. Their positions are shown at instants *1–3*, separated by equal time intervals.

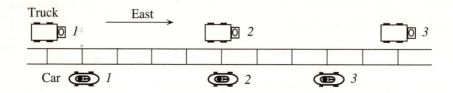

A. Describe the motion of the car and the truck (*i.e.*, the direction of motion of each object and whether it is speeding up, slowing down, or moving with constant speed).

B. Complete the diagram at right by drawing the car and the truck at their positions at instants *2* and *3 as measured in the reference frame of the truck*.

Explain how your completed diagram is consistent with the fact that the truck is at rest in its own frame of reference.

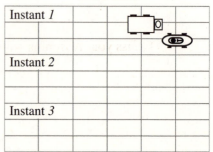

Diagram for the reference frame of the truck

Instant *1*			
Instant *2*			
Instant *3*			

Tutorials in Introductory Physics
McDermott, Shaffer, & P.E.G., U. Wash.

©Prentice Hall, Inc.
First Edition, 2002

C. Use your completed diagram to sketch average velocity vectors for the car *in the reference frame of the truck* for the intervals indicated.

In the reference frame of the truck:

- is the car moving to the *east, moving to the west,* or *at rest?* Explain.

Average velocity of car in frame of truck

from *1* to *2* [   ]

from *2* to *3* [   ]

- is the car *speeding up, slowing down,* or *moving with constant speed?* Explain.

D. During a small time interval Δ*t* from just before to just after instant *2*, does the car *move to the east, move to the west,* or *remain at rest* in the reference frame of the truck? Explain.

In the space provided, draw an arrow to indicate the direction of the *instantaneous* velocity vector of the car in the reference frame of the truck at instant *2*. If the velocity is zero, state that explicitly.

Direction of instantaneous velocity of car in frame of truck at instant *2*

[   ]

Consider the following statements:

Statement 1: *"At instant 2 the car and the truck are side by side, so the velocity of the car in the truck's frame is zero at that instant."*

Statement 2: *"Before instant 2, the truck is catching up to the car, so the truck sees the car as slowing down."*

Do you agree or disagree with each of the statements? Explain.

⇨ Discuss your reasoning with an instructor.

I.   Identifying forces

Two people are attempting to move a large block. The block, however, does not move. Chris is
pushing on the block. Pam is pulling on a rope attached to the block.

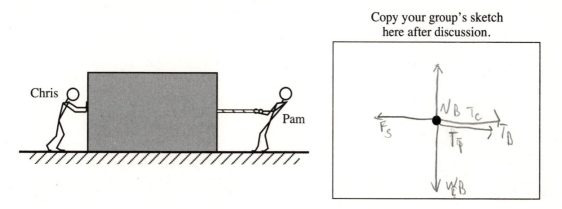

Copy your group's sketch
here after discussion.

A. Draw a large dot on your large sheet of paper to represent the block. Draw vectors with their
"tails" on the dot to show the forces exerted *on* the block. Label each vector and write a brief
description of that force next to the vector.

In Newtonian physics, all forces are considered as arising from an interaction between *two*
objects. Forces are specified by identifying the object *on which* the force is exerted and the object
that is exerting the force. For example, in the situation above, a gravitational force is exerted *on*
the block *by* the Earth.

B. Describe the remaining forces you have indicated above in a similar fashion.

Static Friction   Normal force   Tench force

From Ground   Center of Weight   chis + Pam

The diagram you have drawn is called a *free-body diagram*. A free-body diagram should
show only the forces exerted *on* the object or system of interest (in this case, forces exerted *on*
the block). Check your free-body diagram and, if necessary, modify it accordingly.

A proper free-body diagram should *not* have anything on it except a representation of the
object and the (labeled) forces exerted on that object. A free-body diagram *never* includes
(1) forces exerted by the object of interest on other objects or (2) sketches of other objects that
exert forces on the object of interest.

C. All forces arise from interactions between objects, but the interactions can take different forms.

Which of the forces exerted on the block require *direct contact* between the block and the object exerting the force? Static friction, tension, Normal force

Which of the forces exerted on the block *do not* arise from direct contact between the block and the object exerting the force?
Weight

We will call forces that depend on contact between two objects *contact forces*. We will call forces that do not arise from contact between two objects *non-contact forces*.

D. There are many different types of forces, including: friction ($\vec{f}$), tension ($\vec{T}$), magnetic forces ($\vec{F}^{\text{mag}}$), normal forces ($\vec{N}$), and the gravitational force ($\vec{W}$, for weight). Categorize these forces according to whether they are contact or non-contact forces.

Contact forces	Non-contact forces
friction	Magnetic
tension	weight
Normal	gravitational

E. Consider the following discussion between two students.

Student 1: *"I think the free-body diagram for the block should have a force by Chris, a force by the rope, and a force by Pam."*

Student 2: *"I don't think the diagram should show a force by Pam. People can't exert forces on blocks without touching them."*

With which student, if either, do you agree? Explain your reasoning.
Student 2

It is often useful to label forces in a way that makes clear (1) the type of force, (2) the object on which the force is exerted, and (3) the object exerting the force. For example, the gravitational force exerted *on* the block *by* the Earth might be labeled $\vec{W}_{\text{BE}}$. Your instructor will indicate the notation that you are to use.

F. Label each of the forces on your free-body diagram in part A in the manner described above.

⇨ Do not proceed until a tutorial instructor has checked your free-body diagram.

Tutorials in Introductory Physics
McDermott, Shaffer, & P.E.G., U. Wash.

©Prentice Hall, Inc.
First Edition, 2002

II. Drawing free-body diagrams

A. Sketch a free-body diagram for a book at rest on a level table.
(*Remember:* A proper free-body diagram should not have anything on it except a representation of the book and the forces exerted *on* the book.)

 Book

Make sure the label for each force indicates:

- the type of force (gravitational, frictional, *etc.*),
- the object on which the force is exerted, and
- the object exerting the force.

1. What evidence do you have for the existence of each of the forces on your diagram?

   The book isn't going through the table

2. What observation can you make that allows you to determine the relative magnitudes of the forces acting on the book?

   They are equal because it is at rest

   How did you show the relative magnitudes of the forces on your diagram?

   Same lengths

B. A second book of greater mass is placed on top of the first.

   Sketch a free-body diagram for each of the books in the space below. Label all the forces as in part A.

   Upper book
   Lower book

Free-body diagram for upper book	Free-body diagram for lower book

Specify which of the forces are contact forces and which are non-contact.

weight

1. Examine all the forces on the two free-body diagrams you just drew. Explain why a force that appears on one diagram *should not* appear on the other diagram.

    The weight on B, because B acts on ...

2. What *type* of force does the upper book exert on the lower book (*e.g.,* frictional, gravitational)?

    Normal

    Why would it be *incorrect* to say that the weight of the upper book acts on the lower book?

    Because it acts on the earth

3. What observation can you make that allows you to determine the relative magnitudes of the forces on the *upper* book?

    Because weight isn't a contact force not flying out

4. Are there any forces acting on the *lower* book that have the same magnitude as a force acting on the *upper* book? Explain.

    From book to book normal force is the same

C. Compare the free-body diagram for the lower book to the free-body diagram for the same book in part A (*i.e.,* before the upper book was added).

    Which of the forces changed when the upper book was added and which remained the same?

    Normal force changed

As discussed earlier, we think of each force acting on an object as being exerted by another object. The first object exerts a force of equal magnitude and opposite direction on the second object. The two forces together are called an *action-reaction* or *Newton's third law* force pair.

D. Which, if any, Newton's third law force pairs are shown in the diagrams you have drawn? On which object does each of the forces in the pair act?

    Normal forces on book1, book 2

Identify any third law force pairs on your diagrams by placing one or more small "✕" symbols through each member of the pair. For example, if you have two sets of third law force pairs shown on your diagrams, mark *each* member of the first pair as ⟶✕⟶ , and each member of the second pair as ⟶✕✕⟶ .

Tutorials in Introductory Physics   ©Prentice Hall, Inc.

McDermott, Shaffer, & P.E.G., U. Wash.   First Edition, 2002

III. Supplement: Contact and non-contact forces

A. A magnet is supported by another magnet as shown at right.

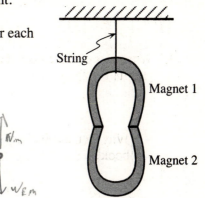

1. Draw a free-body diagram for magnet 2. The label for each
    of the forces on your diagram should indicate:

    • the type of force (*e.g.,* gravitational, normal),
    • the object on which the force is exerted, and
    • the object exerting the force.

2. Suppose that the magnets were replaced by stronger magnets of the same mass.

    If this changes the free-body diagram for magnet 2, sketch the new free-body diagram
    and describe how the diagram changes. (Label the forces as you did in part 1 above.) If
    the free-body diagram for magnet 2 does not change, explain why it does not.

    Normal force is greater

3. Can a magnet exert a non-contact force on another object?

    Yes

    Can a magnet exert a contact force on another object?

    Yes

    Describe how you can use a magnet to exert *both* a contact force and a non-contact force
    on another magnet.

    Yes   N S ~ N S

4. To ensure that you have accounted for all the forces acting on magnet 2 in parts 1 and 2:

    List all the non-contact forces acting on magnet 2.

    Weight
    Magnetic

    List all the contact forces acting on magnet 2. (*Hint:* Which objects are in *contact*
    with magnet 2?)

    Magnetic
    Normal

B. An iron rod is held up by a magnet as shown. The magnet is held up
by a string.

1. In the spaces below, sketch a free-body diagram for the iron rod
and a separate free-body diagram for the magnet.

The label for each of the forces on your diagrams should
indicate:

• the type of force (*e.g.,* gravitational, normal),
• the object on which the force is exerted, and
• the object exerting the force.

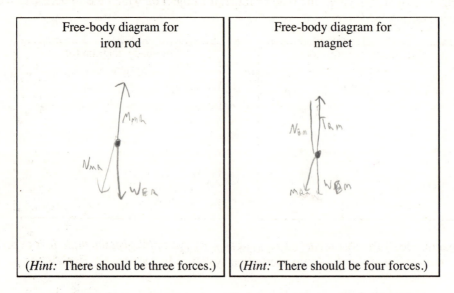

Free-body diagram for iron rod	Free-body diagram for magnet
(*Hint:* There should be three forces.)	(*Hint:* There should be four forces.)

2. For each of the forces shown in your diagram for the iron rod, identify the corresponding
force that completes the Newton's third law (or action-reaction) force pair.

3. How would your diagram for the iron rod change if the magnet were replaced with a
stronger magnet? Which forces would change (in type or in magnitude)? Which forces
would remain the same?

all forces would be in the same direction and

magnetism + Normal force would change

I. Applying Newton's laws to interacting objects: constant speed

Three identical bricks are pushed across a table at *constant speed* as shown. The hand pushes horizontally. (*Note:* There is friction between the bricks and the table.)

Call the stack of two bricks system A and the single brick system B.

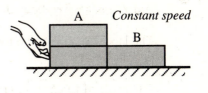

A. Compare the *net force* (magnitude and direction) on system A to that on system B. Explain how you arrived at your comparison.

B. Draw separate free-body diagrams for system A and system B. Label each of the forces in your diagrams by identifying: the type of force, the object on which the force is exerted, and the object exerting the force.

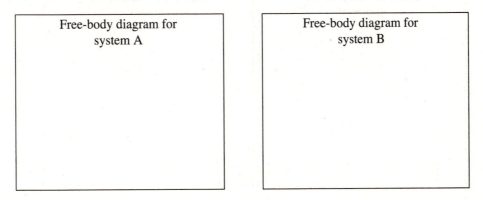

C. Is the magnitude of the force exerted on system A by system B *greater than, less than,* or *equal to* the magnitude of the force exerted on system B by system A? Explain.

Would your answer change if the hand were pushing system B to the left instead of pushing system A to the right? If so, how? If not, why not?

D. Identify all the *Newton's third law (action-reaction)* force pairs in your diagrams by placing one or more small "✕" symbols through each member of the pair *(i.e.,* mark each member of the first pair as →✕→ , each member of the second pair as →✕✕→ , *etc.).*

What criteria did you use to identify the force pair(s)?

Is your answer to part C consistent with your identification of Newton's third law (or action-reaction) force pairs? If so, explain how it is consistent. If not, resolve the inconsistency.

Tutorials in Introductory Physics
McDermott, Shaffer, & P.E.G., U. Wash.

©Prentice Hall, Inc.
First Edition, 2002

E. Rank the magnitudes of all the *horizontal* forces that you identified on your free-body diagrams in part B. (*Hint:* Recall that the bricks are pushed so that they move at constant speed.)

Did you apply Newton's second law in comparing the magnitudes of the horizontal forces? If so, how?

Did you apply Newton's third law in comparing the magnitudes of the horizontal forces? If so, how?

What information besides Newton's laws did you need to apply in comparing the magnitudes of the horizontal forces?

F. Suppose the mass of each brick is 2.5 kg, the coefficient of kinetic friction between the bricks and the table is 0.2, and the bricks are moving at a constant speed of 0.50 m/s.

Determine the magnitude of each of the forces that you drew on your free-body diagrams in part B. (Use the approximation $g = 10$ m/s^2.)

Would your answers change if the bricks were moving half as fast? If so, how? If not, why not?

⇨ Discuss your answers to section I with a tutorial instructor before continuing.

Tutorials in Introductory Physics
McDermott, Shaffer, & P.E.G., U. Wash.

©Prentice Hall, Inc.
First Edition, 2002

II. Applying Newton's laws to interacting objects: varying speed

Suppose the bricks were pushed by the hand with the same force as in section I; however, the coefficient of kinetic friction between the bricks and the table is *less than* that in section I.

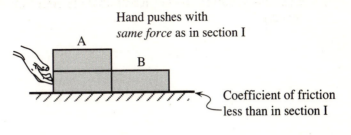

Hand pushes with *same force* as in section I

A

B

Coefficient of friction less than in section I

A. Describe the motions of systems A and B. How does the motion compare to that in part I?

B. Compare the *net force* (magnitude and direction) on system A to that on system B. Explain.

C. Draw and label separate free-body diagrams for systems A and B.

Free-body diagram for system A	Free-body diagram for system B

D. Consider the following discussion between two students.

Student 1: *"System A and system B are pushed by the same force as before, so they will have the same motion as in section I."*

Student 2: *"I disagree. I think that they are speeding up since friction is less. So now system A is pushing on system B with a greater force than system B is pushing on system A."*

With which student, if either, do you agree? Explain your reasoning.

E. Rank the magnitudes of all the *horizontal* forces that appear on your free-body diagrams in part C. Explain your reasoning. (Describe explicitly how you used Newton's second and third laws to compare the magnitudes of the forces.)

Is it possible to *completely* rank the horizontal forces in this case?

Tutorials in Introductory Physics
McDermott, Shaffer, & P.E.G., U. Wash.

©Prentice Hall, Inc.
First Edition, 2002

III. Applying Newton's laws to a system of interacting objects

Let C represent the system consisting of all three bricks. The
motion of the bricks is the same as in section II.

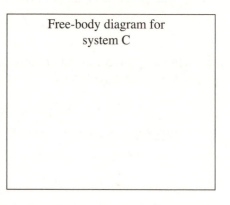

System C

A. Compare the magnitude of the *net force* on system C to
the magnitudes of the *net forces* on systems A and B.
Explain.

B. Draw and label a free-body diagram for system C.

Compare the forces that appear on your free-body
diagram for system C to those that appear on your
diagrams for systems A and B in section II.

For each of the forces that appear on your diagram
for system C, list the corresponding force (or forces)
on your diagrams for systems A and B.

Free-body diagram for
system C

Are there any forces on your diagrams for systems A and B that you did not list? If so, what
characteristic do these forces have in common that none of the others share?

Why is it not necessary to consider these forces in determining the motion of system C?

Note that such forces are sometimes called *internal forces*, to be distinguished from *external*

forces.

IV. Interpreting free-body diagrams

At right is a free-body diagram for a cart. All forces have been
drawn to scale.

In the space below, sketch the cart, rope, *etc.*, as they would
appear in the laboratory.

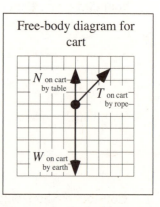

Free-body diagram for
cart

What can you say about the motion of the cart based on the free-body diagram? For example,
could the cart be: moving to the left? moving to the right? stationary? Explain whether each
case is possible and, if so, describe the motion of the cart.

Tutorials in Introductory Physics
McDermott, Shaffer, & P.E.G., U. Wash.

©Prentice Hall, Inc.
First Edition, 2002

I. Blocks connected by a rope

Two blocks, A and B, are tied together with a
rope of mass M. Block B is being pushed with a
constant horizontal force as shown at right.
Assume that there is no friction between the
blocks and the table and that the blocks have
already been moving for a while at the instant
shown.

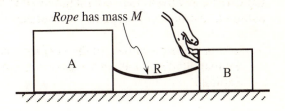

Rope has mass M

A. Describe the motions of block A, block B, and the rope.

B. On a large sheet of paper, draw a separate free-body diagram for each block and for the rope.
    Clearly label the forces.

Copy your free-body diagrams below after discussion.

Free-body diagram for block A	Free-body diagram for rope	Free-body diagram for block B

C. Identify all the *Newton's third law (action-reaction)* force pairs in your diagrams by placing
    one or more small "×" symbols through each member of the pair *(i.e.,* mark each member of
    the first pair as →×→ , each member of the second pair as →××→ , *etc.).*

D. Rank, from largest to smallest, the magnitudes of the *horizontal components* of the forces on
    your diagrams. Explain your reasoning.

E. Consider the horizontal components of the forces exerted *on the rope* by blocks A and B. Is
    your answer above for the relative magnitudes of these components consistent with your
    knowledge of the net force on the rope?

⇨ Check your reasoning with a tutorial instructor before proceeding.

Tutorials in Introductory Physics
McDermott, Shaffer, & P.E.G., U. Wash.

II. Blocks connected by a very light string

The blocks in section I are now connected with a very light, flexible, and inextensible string of mass m $(m < M)$.

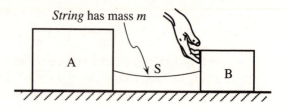

String has mass m

A

S

B

A. If the motion of the blocks is the same as in section I, how does the net force on the *string* compare to the net force on the *rope?*

   1. Determine whether the net force on each of the objects is *greater than, less than,* or *equal to* the net force on the object in section I. Explain.

- block A

- block B

- the system composed of the blocks and the connecting rope or string

   2. Compare the horizontal components of the following pairs of forces:

- the force on the string by block A and the force on the rope by block A. Explain.

- the force on the string by block B and the force on the rope by block B. Explain.

B. Suppose the mass of the string that connects blocks A and B becomes smaller and smaller, but the motion remains the same as in section I. What happens to:

- the magnitude of the net force on that connecting string?

- the magnitudes of the forces exerted on that connecting string by blocks A and B?

C. A string exerts a force on each of the two objects to which it is attached. For a massless string, the magnitude of both forces is often referred to as "the tension in the string."

Justify the use of this approach, in which a *single value* is assumed for the magnitude of both forces.

D. If you know that the net force on a massless string is zero, what, if anything, can you infer about its motion?

Is it possible to exert a non-zero force on a massless string? Is it possible for a massless string to have a non-zero *net* force? Explain.

⇨ Discuss your answers above with a tutorial instructor before continuing.

III. The Atwood's machine

The Atwood's machine at right consists of two identical objects connected by a massless string that runs over an ideal pulley. Object B is initially held so that it is above object A and so that neither object can move.

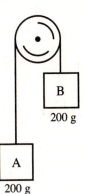

A. Predict the subsequent motions of objects A and B after they are released. Explain the basis for your description. Do not use algebra.

B. Draw separate free-body diagrams for objects A and B. Are your free-body diagrams consistent with your prediction of the motion of the objects?

Object B is replaced by object C, of greater mass. Object C is initially held so that it is higher than object A and so that neither object can move.

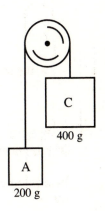

C. *Predict:*

• what will happen to object C when it is released.

• how the motion of object C will compare to the motion of object A after they are released.

Explain the basis for your predictions. Do not use algebra.

Tutorials in Introductory Physics
McDermott, Shaffer, & P.E.G., U. Wash.

©Prentice Hall, Inc.
First Edition, 2002

D. Draw and label separate free-body diagrams for objects A and C *after* they are released. Indicate the relative magnitudes of the forces by the relative lengths of the force vectors.

Are the predictions you made in part C consistent with your free-body diagrams for objects A and C? If so, explain why they are consistent. If not, then resolve the inconsistency.

E. The weight of a 200 g mass has magnitude $(0.2 \text{ kg})(9.8 \text{ m/s}^2) \approx 2 \text{ N}$. Similarly, the weight of a 400 g mass is approximately 4 N in magnitude.

   1. How does the force exerted on object A by the string compare to these two weights?

   2. How does the force exerted on object C by the string compare to these two weights?

   Explain your answers.

   3. How does the net force on object A compare to the net force on object C? Explain.

F. Consider the following statement about the Atwood's machine made by a student.

   "All strings can do is transmit forces from other objects. That means that the string in the Atwood's machine just transmits the weight of one block to the other."

   Do you agree with this student? Explain your reasoning.

Tutorials in Introductory Physics
McDermott, Shaffer, & P.E.G., U. Wash.

©Prentice Hall, Inc.
First Edition, 2002

I. Relating work and changes in kinetic energy

A. A block is moving to the left on a frictionless, horizontal table. A hand exerts a constant horizontal force on the block.

1. Suppose that the work done on the block by the hand is positive. Draw arrows at right to show the direction of the displacement of the block and the direction of the force by the hand.

Displacement of block	Force on block by hand

Explain how you chose the direction of the force on the block by the hand.

Is the block *speeding up, slowing down,* or *moving with constant speed?* Explain.

2. Suppose that the block again moves to the left but now the work done by the hand is negative. In the space at right, draw arrows to represent the direction of the displacement of the block and the direction of the force by the hand.

Displacement of block	Force on block by hand

Explain how you chose the direction of the force on the block by the hand.

Is the block *speeding up, slowing down,* or *moving with constant speed?* Explain.

B. In a separate experiment, two hands push horizontally on the block. Hand 1 does positive work and hand 2 does negative work ($W_{H_1} > 0$; $W_{H_2} < 0$).

For each of the following cases, draw a free-body diagram for the block that shows all the horizontal and vertical forces exerted on the block. Tell whether the sum, $W_{H_1} + W_{H_2}$, is *positive, negative,* or *zero.*

$$W_{H_1} > 0$$
$$W_{H_2} < 0$$

The block moves to the right and speeds up.	The block moves to the left and speeds up.
The block moves to the right and slows down.	The block moves to the left with constant speed.

Tutorials in Introductory Physics
McDermott, Shaffer, & P.E.G., U. Wash.

©Prentice Hall, Inc.
First Edition, 2002

C. Shown at right is a side-view diagram of the displacement, $\Delta\vec{s}_o$, that a block undergoes on a table when pushed by a hand. The horizontal force on the block by the hand, $\vec{F}_{BH}$, is also shown.

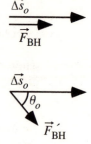

Side view

1. Suppose instead that a hand pushes with a force of the same magnitude, F_{BH}, as before but now at an angle below the horizontal, as shown in the side-view diagram at right. Is the work done by the new force *greater than*, *less than*, or *equal to* the work done by the original force?

   Explain how you used the definition of work to obtain your answer.

2. Suppose instead that a hand pushes with a force of the same magnitude, F_{BH}, as before but instead does zero work. In the space at right, draw an arrow to represent the direction of the force by the hand in this case.

Displacement of block	Force on block by hand
→	

D. Recall the motion of the block in part B. For each force that you identified, state whether that force did *positive work*, *negative work*, or *zero work* on the block. Explain.

The sum of the works done by *all* forces exerted on an object is called the *net work*, W_{net}. Is the *net work* done on the block in part B *positive*, *negative*, or *zero*? Base your answer on your free-body diagram and your knowledge of the block's motion.

Is the net work done on the block *greater than*, *less than*, or *equal to* the work done by the net force on the block? Explain your reasoning.

E. Generalize from your answers to parts A–D to describe how the speed of an object changes if the net work done on the object is (1) positive, (2) negative, or (3) zero.

Discuss how your results are consistent with the work-energy theorem discussed in class ($W_{net} = \Delta K = K_{final} - K_{initial}$).

⇨ Check your answers with a tutorial instructor before proceeding.

Tutorials in Introductory Physics
McDermott, Shaffer, & P.E.G., U. Wash.

©Prentice Hall, Inc.
First Edition, 2002

II. Applying the work-energy theorem

Base your answers below on the work-energy theorem and your results from section I.

A. A glider, glider A, is pulled by a string across a level, frictionless table. The string exerts a constant horizontal force.

    1. How does the net work done on the glider in moving through a distance $2d$ compare to the net work done on the glider in moving through a distance d?

       Assume that the glider starts from rest. Find the ratio of the speed after the glider has moved a distance $2d$ to the speed of the glider after moving a distance d. Explain.

    2. A string pulls a second glider, glider B, across a frictionless table. The string exerts the same force on glider B as did the string on glider A. The mass of glider B is greater than that of glider A $(m_B > m_A)$. Both gliders start from rest.

       After each glider has been pulled a distance d, is the kinetic energy of glider A *greater than*, *less than*, or *equal to* the kinetic energy of glider B? Explain.

B. The diagrams at right show two identical gliders that move to the right *without friction*. The hands exert identical, horizontal forces on the gliders. The second glider experiences an additional, smaller force from a massless string held as shown.

    Suppose the gliders move through identical displacements.

    Is the work done on glider 1 by the hand *greater than*, *less than*, or *equal to* the work done on glider 2 by the hand? Explain.

    Is the change in kinetic energy of glider 1 *greater than, less than,* or *equal to* the change in kinetic energy of glider 2? Base your answer on your knowledge of the net work done on each object.

C. A block on a frictionless table is connected to a spring as shown. The spring is initially unstretched. The block is displaced to the *right* of point *R* and is then released.

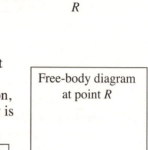

Equilibrium position
(spring unstretched)

Table is
frictionless

L

R

1. When the block passes point *R*, is the spring compressed or stretched?

   Does your answer depend on the direction in which the block is moving? Explain.

2. In the space provided, draw a free-body diagram for the block at the instant the block passes point *R* moving to the left. Draw arrows to represent the directions of the velocity, the acceleration, and the net force on the block, all at that instant. If any quantity is zero, state so explicitly.

   Free-body diagram
   at point *R*

Direction of velocity at point *R*	Direction of net force at point *R*	Direction of acceleration at point *R*

   Is the net work done on the block from the point of release to point *R*, *positive*, *negative*, or *zero*? Explain.

3. At some instant, the block passes point *L* moving to the left. Draw a free-body diagram for the block at that instant. Also, draw arrows to represent the directions of the velocity, the net force, and the acceleration, all at that instant. If any quantity is zero, state so explicitly.

   Free-body diagram
   at point *L*

Direction of velocity at point *L*	Direction of net force at point *L*	Direction of acceleration at point *L*

   During a small displacement of the block from the right of point *L* to the left of point *L:*

   • Is the net work done on the block *positive*, *negative*, or *zero*? Explain.

   • Does the speed of the block *increase, decrease,* or *remain the same?* Explain how your answer is consistent with the work-energy theorem.

I.   Relating forces to changes in kinetic energy and momentum

Two carts, A and B, are initially at rest on a horizontal frictionless table as shown in the top-view diagram below. A constant force of magnitude F_o is exerted on each cart as it travels between two marks on the table. Cart B has a greater mass than cart A.

Top view

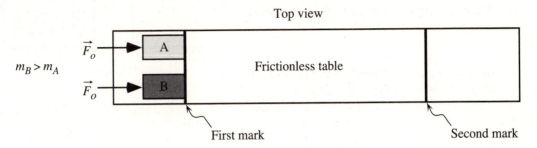

A. Three students discuss the final momentum and kinetic energy of each cart.

Student 1:   *"Since the same force is exerted on both carts, the cart with the smaller mass will move quickly, while the cart with the larger mass will move slowly. The momentum of each cart is equal to its mass times its velocity."*

Student 2:   *"This must mean that the speed compensates for the mass and the two carts have equal final momenta."*

Student 3:   *"I was thinking about the kinetic energies. Since the velocity is squared to get the kinetic energy but mass isn't, the cart with the bigger speed must have more kinetic energy."*

In the space below, write down whether you agree or disagree with the statements made by each student.

B. Which cart takes longer to travel between the two marks? Explain your reasoning.

C. Use Newton's second law and the definition of acceleration to derive an equation for each cart relating the net force on the cart to the change in velocity of the cart ($\Delta \vec{v}_A$ or $\Delta \vec{v}_B$) and the time interval (Δt_A or Δt_B) that the cart spends between the two marks.

1. Is the quantity $m_A |\Delta \vec{v}_A|$ *greater than*, *less than*, or *equal to* $m_B |\Delta \vec{v}_B|$? Explain how you can tell.

Tutorials in Introductory Physics
McDermott, Shaffer, & P.E.G., U. Wash.

©Prentice Hall, Inc.
First Edition, 2002

For a constant net force, the quantity $\vec{F}_{net}\,\Delta t$ is called the *impulse* imparted to the object.

2. Is the magnitude of the impulse imparted to cart A *greater than, less than,* or *equal to* the magnitude of the impulse imparted to cart B? Explain your reasoning.

3. Write an equation showing how the impulse imparted to cart A is related to the *change in momentum vector* of cart A ($\Delta\vec{p}_A$), where momentum, denoted by $\vec{p}$, is the product of the mass and velocity of the object.

This relationship is known as the *impulse-momentum theorem*.

4. Is the magnitude of the final momentum of cart A (p_{A_f}) *greater than, less than,* or *equal to* the magnitude of the final momentum of cart B (p_{B_f})? Explain.

D. How does the net work done on cart A *($W_{net,\,A}$)* compare to the net work done on cart B *($W_{net,\,B}$)*? Explain.

Is the kinetic energy of cart A *greater than, less than,* or *equal to* the kinetic energy of cart B after they have passed the second mark?

E. Refer again to the discussion among the three students in part A. Do you agree with your original answer?

If you disagree with any of the students, identify what is incorrect with their statements.

⇨ Discuss your answers to parts C and D with a tutorial instructor before continuing.

Tutorials in Introductory Physics
McDermott, Shaffer, & P.E.G., U. Wash.

II. Applying the work-energy and impulse-momentum theorems

Obtain a wedge, a ball, a cardboard ramp, and enlargements of the two diagrams below (or sketch them on a large sheet of paper).

(*Note:* It is important that each time the ball is rolled it has the same speed on the level region I. Place a mark halfway up the wedge and release the ball from the mark each time.) Ignore friction and the rotation of the ball.

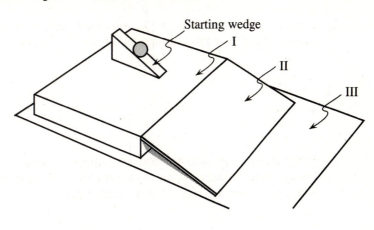

A. Release the ball so that it rolls straight toward the ramp (motion 1).

Observe the motion of the ball.

Sketch the trajectory of the ball on an enlargement for motion 1.

On the enlargement, draw arrows to show the directions of (1) the acceleration of the ball and (2) the net force on the ball while it is on the ramp (*i.e.,* in region II).

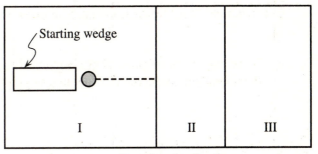

Top view, motion 1

B. Release the ball at an angle to the ramp as shown (motion 2).

Observe the motion of the ball.

Sketch the trajectory of the ball on an enlargement for motion 2.

On the enlargement, draw arrows to show the directions of (1) the acceleration of the ball and (2) the net force on the ball while it is on the ramp (*i.e.,* in region II).

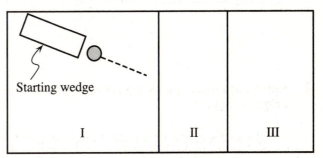

Top view, motion 2

Tutorials in Introductory Physics
McDermott, Shaffer, & P.E.G., U. Wash.

©Prentice Hall, Inc.
First Edition, 2002

C. How does the direction of the net force on the ball in motion 2 compare to the direction of the net force on the ball in motion 1? Explain.

Is the direction of the acceleration of the ball in motion 2 consistent with the fact that the ball speeds up and its trajectory curves? Explain.

D. How does the change in kinetic energy of the ball in motion 1 compare to the change in kinetic energy of the ball in motion 2?

1. Is your answer consistent with the net work done on the ball in motions 1 and 2? Explain.

2. How does the final speed of the ball in motion 1 compare to the final speed in motion 2? Explain.

E. For motion 1, draw vectors in region II of the enlargement that represent the momentum of the ball at the top of the ramp and at the bottom of the ramp (*i.e.*, at the top and bottom of region II). Use these vectors to construct the change in momentum vector $\Delta \vec{p}$.

How is the direction of $\Delta \vec{p}$ related to the direction of the net force on the ball as it rolls down the ramp? Is your answer consistent with the impulse-momentum theorem?

Tutorials in Introductory Physics
McDermott, Shaffer, & P.E.G., U. Wash.

©Prentice Hall, Inc.
First Edition, 2002

F. For motion 2, draw vectors in region II of the enlargement that represent the initial and the final momentum of the ball. Draw these vectors using the same scale that you used for motion 1 (*i.e.,* the relative lengths should represent the relative magnitudes). Use these vectors to construct the change in momentum vector $\Delta\vec{p}$ for motion 2.

How should the direction of $\Delta\vec{p}$ compare to the direction of the net force on the ball as it rolls down the ramp? If necessary, modify your diagram to be consistent with the impulse-momentum theorem.

G. Consider the change in momentum vectors you constructed for motions 1 and 2.

1. How do they compare in direction? How do they compare in magnitude?

2. On the basis of your answer above, compare the time that the ball spends on the ramp in motion 1 to the time it spends on the ramp in motion 2. Explain. (*Hint:* Can you use the impulse-momentum theorem to compare the time intervals?)

Is your answer consistent with the trajectory of the ball in each motion? Explain.

I.   Analyzing collisions

Two experiments are conducted with gliders on a level, *frictionless* track.

In the first experiment, glider A is launched toward a stationary glider, glider M. After the collision, glider A has reversed direction.

In the second experiment, glider M is replaced by glider N, which has the same mass as glider M. Glider A has the same initial velocity as in experiment 1. After the collision, glider A is at rest.

The mass of glider A is one-fifth that of glider M and glider N *(i.e., $m_M = 5m_A$, $m_M = m_N$).*

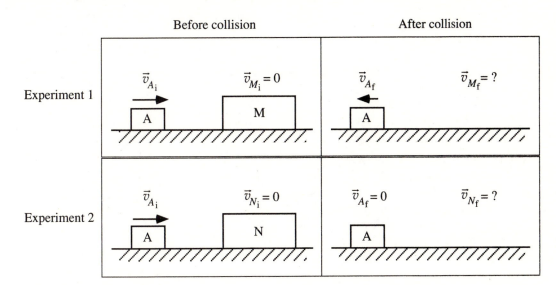

A. What differences between gliders M and N could account for their differences in behavior during the collisions?

B. For experiment 1, draw and label separate free-body diagrams for glider A and glider M at an instant during the collision (*i.e.,* while the gliders are in contact).

Free-body diagram for glider A	Free-body diagram for glider M

1. While the gliders are in contact, how does the net force on glider A compare to the net force on glider M? Discuss both magnitude and direction.

How, if at all, would this comparison differ if you had chosen a different instant (while the gliders are still in contact)? Explain.

Tutorials in Introductory Physics
McDermott, Shaffer, & P.E.G., U. Wash.

©Prentice Hall, Inc.
First Edition, 2002

2. Consider the small time interval (Δt_o) while the gliders are still in contact. For the two gliders, how does the product $\vec{F}_{net,A}\Delta t_o$ compare to the product $\vec{F}_{net,M}\Delta t_o$? Discuss both magnitude and direction. Explain.

Apply Newton's second law ($\vec{F}_{net} = m\dfrac{\Delta\vec{v}}{\Delta t}$) to each of the colliding gliders in Experiment 1 to compare the *change in momentum* ($\Delta\vec{p} = m\Delta\vec{v}$) of gliders A and M during the collision. Discuss both magnitude and direction. Explain.

C. In the spaces provided, draw and label vectors to represent the *initial momentum*, the *final momentum*, and the *change in momentum* of glider A in each experiment.

$\vec{p}_i, \vec{p}_f$, and $\Delta\vec{p}$ for glider A Experiment 1	$\vec{p}_i, \vec{p}_f$, and $\Delta\vec{p}$ for glider A Experiment 2

1. Is the magnitude of the *change in momentum* of glider A in experiment 1 *greater than, less than,* or *equal to* the magnitude of the *change in momentum* of glider A in experiment 2? Explain.

2. Is the magnitude of the *change in momentum* of glider M in experiment 1 *greater than, less than,* or *equal to* the magnitude of the *change in momentum* of glider N in experiment 2? Explain.

After the collisions, is the speed of glider M *greater than, less than,* or *equal to* the speed of glider N? Explain.

D. A student compares the final speeds of gliders M and N.

"In experiment 2, glider A transfers all of its momentum to glider N, whereas in experiment 1, glider A still has some momentum left so glider M does not get as much. Therefore, glider N has a greater final speed than glider M."

Do you agree or disagree with this statement? Explain.

⇨ Discuss your answers with a tutorial instructor.

Tutorials in Introductory Physics
McDermott, Shaffer, & P.E.G., U. Wash.

II.   Applying momentum conservation to systems of multiple objects

An experiment is conducted on a frictionless air track
in which a glider, glider C, is launched toward a
second glider, glider D.

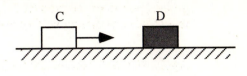

A.   Suppose that glider D is free to move and glider C
rebounds.

1.   In the spaces provided, draw separate free-body diagrams for each glider and for the
system of the two gliders, system S, at an instant during the collision.

Free-body diagram for glider C	Free-body diagram for glider D	Free-body diagram for system S

Which forces in your free-body diagrams for glider C and glider D do not have
corresponding forces on the free-body diagram for system S?

2.   The *momentum of a system* containing multiple objects can be defined to be the sum of
the momenta of the constituent objects.

Use this definition to write an expression for the *change in momentum* of system S in
terms of the *change in momentum* of glider C and of glider D.

3.   Does the momentum of each of the following change during the collision? Explain how
you can tell.

•   glider C

•   glider D

•   system S

Are your answers consistent with your free-body diagrams and the direction of the net
force in each case? If not, resolve any inconsistencies.

4.   How, if at all, would your answer about the change in momentum of system S differ if
glider D were replaced by a much more massive glider? Explain.

B. A second experiment is performed in which glider D is fixed in place. Glider C is launched toward glider D with the same velocity as in the first experiment, and it rebounds with the same speed that it had initially.

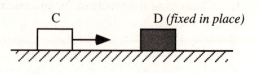

1. In the spaces provided, draw separate free-body diagrams for each glider and for the system of the two gliders at an instant during the collision in this second experiment.

Free-body diagram for glider C	Free-body diagram for glider D	Free-body diagram for system S

Explain how the fact that glider D is fixed in place is reflected in your free-body diagrams.

2. Does the momentum of each of the following change during the collision? Explain.

- glider C

- glider D

- system S

C. Consider the two experiments described above. When the momentum of an object or system of objects did not change:

- were *external* forces exerted on the object or system?

- was there a *net* force on the object or system?

D. When the momentum of an object or system of objects does not change with time, the momentum of the object or system is said to be *conserved*.

On the basis of your results above, describe the circumstances under which the momentum of an object or system of objects is conserved.

E. Two students discuss the second experiment, in which glider D is fixed in place.

Student 1:   *"When one object hits another, the momentum of the system is always conserved."*

Student 2:   *"That's right, the momentum of glider C is the same before and after the collision."*

Describe the *error* in each student's statement.

I. Changes in momentum for interacting objects

Two blocks connected by a massless spring are on top of a frictionless, level table. The blocks are pulled apart slightly so that the spring is stretched, and while they are held apart they are given identical initial velocities in a direction perpendicular to the spring. The blocks are then released at the same time.

The mass of block A is two and a half times the mass of block B.

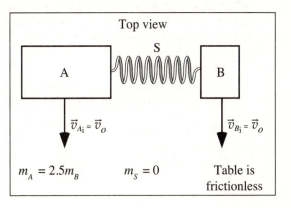

A. Draw separate free-body diagrams for each block and for the spring immediately after release. Indicate separately the *vertical* forces (perpendicular to the table top) and the *horizontal* forces (parallel to the table top). Clearly label all forces.

Free-body diagram for block A	Free-body diagram for spring	Free-body diagram for block B
Vertical forces	Vertical forces	Vertical forces
Horizontal forces	Horizontal forces	Horizontal forces

Identify all the *Newton's third law (action-reaction)* force pairs in your diagrams by placing one or more small "×" symbols through each member of the pair *(i.e.,* mark each member of the first pair as ⟶×➤, each member of the second pair as ⟶××➤, *etc.).*

B. Rank the magnitudes of all the *horizontal* forces on your diagrams. If any of the horizontal forces have the same magnitude, state that explicitly. Explain how Newton's second and third laws can be used to arrive at this ranking.

How does the net force on block A compare to the net force on block B? Discuss both magnitude and direction.

Does this comparison of the net forces hold true for all times following the release of the blocks? Explain your reasoning.

Tutorials in Introductory Physics
McDermott, Shaffer, & P.E.G., U. Wash.

©Prentice Hall, Inc.
First Edition, 2002

C. The velocity vectors for blocks A and B are shown for a time immediately before release. Draw a change in velocity vector for each block for a small time interval Δt after release.

$\vec{v}_{A_i}$	$\vec{v}_{B_i}$	$\Delta\vec{v}_A$	$\Delta\vec{v}_B$

Explain how Newton's second law and the definition of acceleration can be used to determine the directions of the change in velocity vectors.

By what factor is the magnitude of $\Delta\vec{v}_B$ greater than the magnitude of $\Delta\vec{v}_A$? Explain.

How does $m_B\Delta\vec{v}_B$ compare to $m_A\Delta\vec{v}_A$ for this small time interval?

Would this comparison change if we considered:

- another interval of time, equally small, occurring much later?

- a much larger time interval?

Explain.

D. Use your knowledge of the velocities and changes in velocities to construct *momentum vectors* and *change in momentum vectors* for the blocks. Also draw a final momentum vector for each block corresponding to the same small time interval as in part C. Show the correct relative magnitudes.

$\vec{p}_{A_i}$	$\vec{p}_{B_i}$	$\Delta\vec{p}_A$	$\Delta\vec{p}_B$	$\vec{p}_{A_f}$	$\vec{p}_{B_f}$

Explain how you determined these vectors.

How would $\Delta\vec{p}_A$ compare to $\Delta\vec{p}_B$ if we considered a much larger time interval? Explain.

Tutorials in Introductory Physics
McDermott, Shaffer, & P.E.G., U. Wash.

©Prentice Hall, Inc.
First Edition, 2002

II. Changes in momentum for systems of interacting objects

Let system C denote the combined system of blocks A and B and the spring S.

A. Draw and label a free-body diagram for system C at a time following the release of the blocks. Indicate separately the *vertical* forces (perpendicular to the table top) and the *horizontal* forces (parallel to the table top). Show separately the net force on system C.

> Which forces in your free-body diagrams in section I are *internal forces* for system C?

Free-body diagram for system C	Net force on system C
Vertical forces	
Horizontal forces	

B. Write an equation for the momentum of system C in terms of the momenta of its constituent bodies.

Compare the momentum of system C *immediately* after the blocks are released to its momentum at the following times:

- a short time later, when the blocks have undergone the change in momentum indicated in section I, and

- a much longer time later.

Explain how you determined your answers.

C. Generalize from your results to answer the following question: Under what condition will the momentum of a system be conserved?

D. Imagine a single object whose mass is equal to the mass of system C and whose momentum is equal at all times to $\vec{p}_C$. Draw an arrow that represents the direction of the velocity of that object. If the velocity is zero, state that explicitly.

Direction of $\vec{v}$

The velocity that you have found is called the velocity of the *center of mass*, $\vec{v}_{cm}$, of system C.

III. Application of momentum conservation to two-dimensional collisions

The spring formerly connecting blocks A and B is disconnected from block B. The blocks are given initial velocities in the directions shown so that they will collide with the spring between them. As in section II, system C refers to the combination of both blocks and spring.

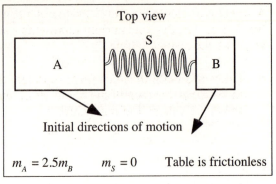

Top view

Initial directions of motion

$m_A = 2.5m_B$ $\quad m_S = 0$ $\quad$ Table is frictionless

A. What are the *external* forces exerted on system C during the collision?

What is the *net force* on system C?

B. The momentum vectors of each block before the collision and of block B after the collision are shown. Complete the table to show the momentum vectors of system C before and after the collision and of block A after the collision.

	Block A	Block B	System C
Momentum *before* collision			
Momentum *after* collision			
Change in momentum			

How do the final speeds of the blocks compare? Explain.

C. Draw arrows that represent the direction of the velocity of the center of mass of system C *before* and *after* the collision.

As a result of the collision, does the speed of the center of mass of system C *increase, decrease,* or *stay the same*? Explain.

Direction of $\vec{v}_{cm_i}$	Direction of $\vec{v}_{cm_f}$

I.   Motion with constant angular velocity

A wheel is spinning *counterclockwise* at a constant rate about a fixed axis. The diagram at right represents a snapshot of the wheel at one instant in time.

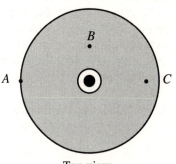

Top view
Wheel spins *counterclockwise*

A. Draw arrows on the diagram to represent the direction of the velocity for each of the points *A*, *B*, and *C* at the instant shown. Explain your reasoning.

    Is the time taken by points *B* and *C* to move through one complete circle *greater than*, *less than*, or *the same as* the time taken by point *A*?

    On the basis of your answer above, determine how the speeds of points *A*, *B*, and *C* compare. Explain.

B. Mark the position of each of the labeled points at a later time when the wheel has completed one half of a turn. Sketch a velocity vector at each point.

    For each labeled point, how does the velocity compare to the velocity at the earlier time in part A? Discuss both magnitude and direction.

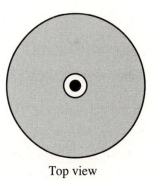

Top view
Wheel spins *counterclockwise*

    Is there one single *linear velocity vector* that applies to every point on the wheel at all times? Explain.

C. Suppose the wheel makes one complete revolution in 2 seconds.

   1. For each of the following points, find the change in angle ($\Delta\theta$) of the position vector during one second. (*i.e.*, Find the angle between the initial and final position vectors.)

      • point *A*

      • point *B*

      • point *C*

   2. Find the rate of change in the angle for any point on the wheel.

The rate you calculated above is called the *angular speed* of the wheel, or equivalently, the magnitude of the *angular velocity* of the wheel. The angular velocity is defined to be a vector that points along the axis of rotation and is conventionally denoted by the symbol $\vec{\omega}$ (the Greek letter *omega*). To determine the direction of the angular velocity vector, we imagine an observer on the axis of rotation who is looking toward the object. If the observer sees the object rotating counterclockwise, the angular velocity vector is directed toward the observer; if the observer sees it rotating clockwise, the angular velocity vector is directed away from the observer.

D. Would two observers on either side of a rotating object agree on the *direction* of the angular velocity vector? Explain.

Would two observers who use different points on an object to determine the angular velocity agree on the *magnitude* of the angular velocity vector? Explain.

E. The diagrams at right show top and side views of the spinning wheel in part A.

On each diagram, draw a vector to represent the angular velocity of the wheel. (Use the convention that ⊙ indicates a vector pointing *out of* the page and ⊗ indicates a vector pointing *into* the page.)

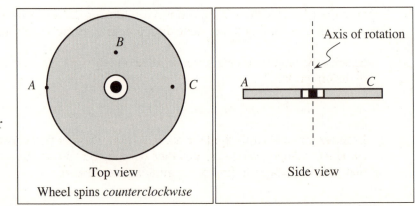

Top view

Wheel spins *counterclockwise*

Side view

Tutorials in Introductory Physics
McDermott, Shaffer, & P.E.G., U. Wash.

©Prentice Hall, Inc.
First Edition, 2002

F. In the space at right sketch the position vectors for point C at the beginning and at the end of a small time interval Δt.

> Sketch of position vectors at
> t_o and $t_o + \Delta t$

1. Label the change in angle ($\Delta\theta$) and the distance between the center of the wheel and point C (r_C). Sketch the path taken by point C during this time interval.

   What is the distance that point C travels during Δt? Express your answer in terms of r_C and $\Delta\theta$.

2. Use your answer above and the definition of linear speed to derive an algebraic expression for the linear speed of point C in terms of the angular speed ω of the wheel.

   What does your equation imply about the relative linear speeds for points farther and farther out on the wheel? Is this consistent with your answer to part A?

II. Motion with changing angular velocity

A. Let $\vec{\omega}_o$ represent the initial angular velocity of a wheel. In each case described below, determine the magnitude of the *change in angular velocity* $\left|\Delta\vec{\omega}\right|$ in terms of $\left|\vec{\omega}_o\right|$.

1. The wheel is made to spin faster, so that eventually, a fixed point on the wheel is going around twice as many times each second. (The axis of rotation is fixed.)

2. The wheel is made to spin at the same rate but in the opposite direction.

B. Suppose the wheel slows down uniformly, so that $\left|\vec{\omega}\right|$ decreases by 8π rad/s every 4 s. (The wheel continues spinning in the same direction and has the same orientation.)

   Specify the angular acceleration $\vec{\alpha}$ of the wheel by giving its magnitude and, relative to $\vec{\omega}$, its direction.

   In linear kinematics we find the acceleration vector by first constructing a *change in velocity vector* $\Delta\vec{v}$ and then dividing that by Δt. Describe the analogous steps that you used above to find the angular acceleration $\vec{\alpha}$.

⇨ Discuss your answers above with a tutorial instructor before continuing.

Tutorials in Introductory Physics
McDermott, Shaffer, & P.E.G., U. Wash.
©Prentice Hall, Inc.
First Edition, 2002

III. Torque and angular acceleration

The rigid bar shown at right is free to rotate about a fixed pivot through its center. The axis of rotation of the bar is perpendicular to the plane of the paper.

$|F| = F_o$

$\theta = 90°$

M

A. A force of magnitude F_o is applied to point M as shown. The force is *always* at a right angle to the bar.

For each of the following cases, determine whether the angular acceleration would be in a *clockwise* or *counterclockwise* sense.

- The bar was initially at rest. *(Hint:* Consider $\Delta\vec{\omega}$.)

- The bar was spinning at a constant rate before the force was applied.

    Does your answer for the angular acceleration depend on whether the bar was originally spinning clockwise or counterclockwise? Explain.

The application point and direction of a force can affect the rotational motion of the object to which the force is applied. The tendency of a particular force to cause an angular acceleration of an object is quantified as the *torque* produced by the force. The torque $\vec{\tau}$ is defined to be the vector cross product $\vec{r} \times \vec{F}$, where $\vec{r}$ is the vector from the axis of rotation to the point where the force is applied. The magnitude of the torque is simply $|\vec{\tau}| = |\vec{r}|\,|\vec{F}|\sin\theta$, where θ is the angle between $\vec{r}$ and $\vec{F}$.

B. Compare the magnitude of the *net torque* about the pivot in part A to that in each case below.

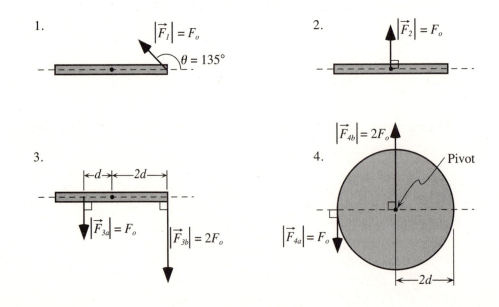

1.

$|\vec{F_1}| = F_o$

$\theta = 135°$

2.

$|\vec{F_2}| = F_o$

3.

$\leftarrow d \rightarrow\leftarrow 2d \rightarrow$

$|\vec{F_{3a}}| = F_o$

$|\vec{F_{3b}}| = 2F_o$

4.

$|\vec{F_{4b}}| = 2F_o$

Pivot

$|\vec{F_{4a}}| = F_o$

$\leftarrow 2d \rightarrow$

Tutorials in Introductory Physics
McDermott, Shaffer, & P.E.G., U. Wash.

©Prentice Hall, Inc.
First Edition, 2002

I. Drawing extended free-body diagrams

A. A ruler is placed on a pivot and held at an angle as shown at right. The pivot passes through the center of the ruler.

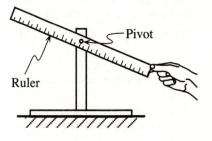

Predict the motion of the ruler after it is released from rest. Explain your reasoning.

Check your prediction by observing the demonstration.

1. Is the angular acceleration of the ruler in a *clockwise sense,* in a *counterclockwise sense,* or *zero?* Explain how you can tell.

   What does your answer imply about the *net torque* on the ruler about the pivot? Explain.

2. What is the direction of the acceleration of the center of mass of the ruler? If $\vec{a}_{cm} = 0$, state that explicitly. Explain how you can tell.

   What does your answer imply about the *net force* acting on the ruler? Explain.

Tutorials in Introductory Physics
McDermott, Shaffer, & P.E.G., U. Wash.

©Prentice Hall, Inc.
First Edition, 2002

B. Draw a free-body diagram for the ruler (after it is released from rest). Draw your vectors on the diagram at right. Draw each force at the point at which it is exerted.

Label each force by identifying:

- the type of force,

- the object on which the force is exerted, and

- the object exerting the force.

The diagram you have drawn is called an *extended free-body diagram*.

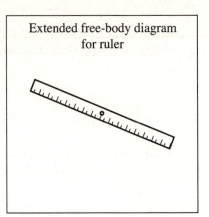

Extended free-body diagram
for ruler

Is the point at which you placed the gravitational force in your diagram consistent with your knowledge of the net torque about the pivot? Explain.

C. How would your free-body diagram change if the ruler were twice its original length and the same mass as before? Explain.

II. Distinguishing the effect of net torque and net force

Two identical spools are held the same height above the floor. The thread from spool A is tied to a support, while spool B is not connected to a support. An "×" is marked on the floor directly below each spool.

Both spools are released from rest at the same instant. (Make the approximation that the thread is massless.)

Draw an extended free-body diagram for each spool at an instant after they are released but before they hit the floor.

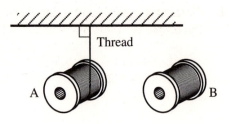

Thread

A   B

Floor

Extended free-body diagram for spool A	Extended free-body diagram for spool B

For each spool, determine the direction of the net torque about the center of the spool. If the net torque is zero, state that explicitly. Explain your reasoning.

Tutorials in Introductory Physics
McDermott, Shaffer, & P.E.G., U. Wash.

©Prentice Hall, Inc.
First Edition, 2002

A. *Predict:*

   • which spool will reach the floor first. Explain how your answer is consistent with your extended free-body diagrams.

   • whether spool A will *strike the floor to the left of the "×," strike the floor to the right of the "×,"* or *fall straight down.* Explain how your answer is consistent with your free-body diagrams.

   Describe how the net force is related to the individual forces on a free-body diagram when the forces are exerted at different points on the object.

B. Obtain two spools and a ring stand. Use the equipment to check your predictions. (Be sure the thread of spool A is vertical before the spools are released.)

   1. How does the *magnitude* of the acceleration of the center of mass $(\vec{a}_{cm})$ of spool A compare to that of spool B? Is this consistent with your free-body diagrams? Explain.

   2. How does the *direction* of the acceleration of the center of mass $(\vec{a}_{cm})$ of spool A compare to that of spool B? Is this consistent with your free-body diagrams? Explain.

C. Consider the following discussion between three students.

Student 1:   *"The string exerts a force that is tangent to the rim of spool A. This force has no component that points toward the center of the spool, so this force does not affect the acceleration of the center of mass."*

Student 2:   *"I disagree. The acceleration of the center of mass of the spool is affected by the string. Any of the force not used up in rotational acceleration will be given to translational acceleration. This is why the acceleration of the center of mass of spool A is less than g."*

Student 3:   *"The net force on spool A is the gravitational force minus the tension force. By Newton's second law, the acceleration of the center of mass is the net force divided by the mass. A force will have the same effect on the motion of the center of mass regardless of whether the force causes rotational motion or not."*

With which student(s), if any, do you agree? Explain your reasoning.

If necessary, revise your description in part A of how the net force is related to forces exerted at different points on an object.

⇨ Discuss your answers with a tutorial instructor before continuing.

D. Write down Newton's second law for each spool. Express your answer in terms of the mass of each spool *(m)*, the acceleration of the center of mass of each spool $(\vec{a}_{cm})$, and the individual forces acting on each spool.

Write down the rotational analogue to Newton's second law for each spool. Express your answer in terms of the relevant rotational quantities, that is, in terms of the angular acceleration $(\vec{\alpha})$, the rotational inertia *(I)*, and the torque $(\vec{\tau}_{net})$. Express the torque in terms of the individual forces and appropriate distances.

Tutorials in Introductory Physics
McDermott, Shaffer, & P.E.G., U. Wash.

©Prentice Hall, Inc.
First Edition, 2002

I. Interpreting center of mass

A. A T-shaped board of uniform mass density has two small holes as shown. Initially, the pivot is placed through the right hole, which corresponds to the *center of mass* of the board. The board is then held in place.

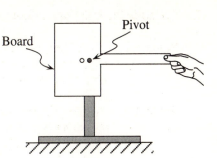

1. *Predict* the motion of the board after it is released from rest. Explain.

2. Check your prediction by observing the demonstration.

   a. Describe the angular acceleration of the board. Explain how you can tell.

   What does your answer imply about the *net torque* about the pivot? Explain.

   b. Describe the acceleration of the center of mass of the board. Explain how you can tell.

   What does your answer imply about the *net force* acting on the board? Explain.

3. Explain how your answers about net torque and net force in question 2 would change, if at all, if there is *appreciable friction* between the board and the pivot and the board remains at rest.

B. Imagine that the board is now hung from the hole to the left of the center of mass.

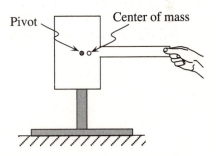

1. *Predict* the subsequent motion of the board after it is released. Explain.

2. On the diagram at right, draw and label an *extended free-body diagram* for the board just after it is released (*i.e.,* for each force, indicate explicitly a single point on the object at which the force can be regarded as acting).

   Explain how the diagram can be used to support your prediction for the motion of the board.

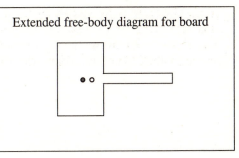

Extended free-body diagram for board

Tutorials in Introductory Physics
McDermott, Shaffer, & P.E.G., U. Wash.

©Prentice Hall, Inc.
First Edition, 2002

3. Obtain a T-shaped board and a pivot. Place the board on the pivot and check your predictions. Resolve any inconsistencies between your predictions and observations.

4. Place the board on the pivot with the center of mass directly above, directly below, and to the left of the pivot. Record your observations below.

An extended body can rotate freely about a fixed pivot if the friction between the object and the pivot is very small. A pivot that has negligible friction is sometimes referred to as a *frictionless pivot*. For all the following exercises and demonstrations in this tutorial, we will assume that the pivot is frictionless.

II. Applying the concept of center of mass

A. Attach clay to the bottom left side of the board so that it remains at rest when placed horizontally on the pivot. (The pivot should still be through the hole used in part B above.)

1. On the figure at right, mark the approximate location of the center of mass of the system composed of clay and board with an "x."

   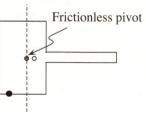

   Is the center of mass of the system located *to the left of, to the right of,* or *along* the vertical line through the pivot? Explain.

2. Suppose that the piece of clay were moved to a new location (point *A*) closer to the pivot.

   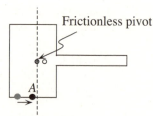

   Predict whether the board would remain in equilibrium. Explain.

   Would the total mass to the left of the pivot change when the clay is moved to point *A?*

   Check your predictions.

Tutorials in Introductory Physics
McDermott, Shaffer, & P.E.G., U. Wash.

©Prentice Hall, Inc.
First Edition, 2002

3. Suppose the piece of clay were moved back to its original location and additional clay were added to it.

    Would the board remain in equilibrium?

    Is there any location along the bottom edge of the board at which this larger piece of clay could be placed so the board will be in equilibrium? If so, is the new location *closer to* or *farther from* the pivot?

    Check your prediction.

4. Generalize your observations from parts 1, 2, and 3:

    * Is it possible to keep the total mass to either side of the pivot unchanged yet change the system so that it is no longer in equilibrium?

    * Is it possible to change the *total mass* to one side of the pivot and still have the system in equilibrium?

    Is it enough to know the total mass to either side of the pivot in order to determine whether the system will be in equilibrium? Explain.

B. A student has balanced a hammer *lengthwise* on a finger. Consider the following dialogue between the student and a classmate.

    Student 1: *"The hammer is balanced because the center of mass is above my finger. The mass is the same on both sides of the center of mass — that is what the center of mass means."*

    Student 2: *"It is not the mass, it is the torque that is the same for both parts of the hammer. If the torques weren't the same, the hammer would rotate."*

    With which statements, if any, do you agree? Explain.

    Explain how one of the students above misinterpreted the term "center of mass."

In part A of section I you observed that the board will remain in equilibrium when placed on the pivot through its center of mass and released from rest. The figure at right shows the board with a vertical line through its center of mass.

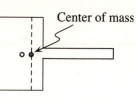

Is the mass of the piece of board to the left of the dashed line *greater than*, *less than*, or *equal to* the mass of the piece of board to the right of the dashed line? Explain.

⇨ Obtain the necessary equipment from a tutorial instructor and check your answers. If necessary, resolve any inconsistencies. (*Hint:* Consider the areas of the pieces of board to the left and to the right of the pivot.)

C. Imagine that the T-shaped board (with no clay attached) were rotated as shown and then released from rest. The pivot passes through the center of mass of the board.

1. *Predict* the subsequent motion of the board.

2. Check your prediction by watching the demonstration. Record your observation below.

3. What does your observation imply about the net torque about the pivot? Explain.

Tutorials in Introductory Physics
McDermott, Shaffer, & P.E.G., U. Wash.

©Prentice Hall, Inc.
First Edition, 2002

Electricity and magnetism

I. Electrical interactions

A. Press a piece of sticky tape, about 15–20 cm in length, firmly onto a smooth unpainted surface, for example, a notebook or an unpainted tabletop. (For ease in handling, make "handles" by folding each end of the tape to form portions that are not sticky.) Then peel the tape off the table and hang it from a support (*e.g.*, a wooden dowel or the edge of a table).

Describe the behavior of the tape as you bring objects toward it (*e.g.*, a hand, a pen).

B. Make another piece of tape as described above. Bring the second tape toward the first. Describe your observations.

It is important, as you perform the experiment above, that you keep your hands and other objects away from the tapes. Explain why this precaution is necessary.

How does the distance between the tapes affect the interaction between them?

C. Each member of your group should press a tape onto the table and write a *"B"* (for bottom) on it. Then press another tape on top of each *B* tape and label it *"T"* (for top).

Pull each pair of tapes off the table as a unit. After they are off the table, separate the *T* and *B* tapes. Hang one of the *T* tapes and one of the *B* tapes from the support at your table.

Describe the interaction between the following pairs of tape when they are brought near one another.

• two *T* tapes   • two *B* tapes   • a *T* and a *B* tape

Tutorials in Introductory Physics
McDermott, Shaffer, & P.E.G., U. Wash.

©Prentice Hall, Inc.
First Edition, 2002

D. Obtain an acrylic rod and a piece of wool or fur. Rub the rod with the wool, and then hold the rod near newly made T and B tapes on the wooden dowel.

Compare the interactions of the rod with the tapes to the interactions between the tapes in part C. Describe any similarities or differences.

We say that the rod and tapes are *electrically charged* when they interact as you have observed.

E. Base your answers to the following questions on the observations you have made thus far.

1. Is it possible that there is only one type of charge? If not, what is the minimum number of different types of charge needed to account for your observations thus far? Explain.

2. By convention, a glass rod is said to be "positively charged" when rubbed with silk. Your instructor will tell you whether your acrylic rod is positively or negatively charged when rubbed with the particular material you are using.

How do two objects that are positively charged interact? Explain how you can tell.

Which tape, T or B, has a positive charge? Explain.

⇨ Discuss part I with a tutorial instructor before continuing.

Please remove all tape from the tabletop before continuing.

II. Superposition

Coulomb's law states that the electric force between two *point charges* acts along the line connecting the two points. (A *point charge* is a charged object that is sufficiently small that the charge can be treated as if it were all located at a single point.) The magnitude of the force on either of the charges is proportional to the product of the charges and is inversely proportional to the square of the distance between the charges.

Tutorials in Introductory Physics
McDermott, Shaffer, & P.E.G., U. Wash.

©Prentice Hall, Inc.
First Edition, 2002

A. Two positive point charges $+q$ and $+Q$ (with $|Q| > |q|$) are held in place a distance s apart.

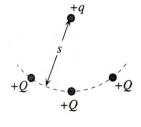

1. Indicate the direction of the electric force exerted on each charge by the other.

2. Is the force on the $+q$ charge by the $+Q$ charge *greater than, less than,* or *equal to* the force on the $+Q$ charge by the $+q$ charge? Explain.

3. By what factor would the magnitude of the electric force on the $+q$ charge change if the charges were instead separated by a distance $2s$?

B. Two more $+Q$ charges are held in place the same distance s away from the $+q$ charge as shown. Consider the following student dialogue concerning the net force on the $+q$ charge:

Student 1: *"The net electric force on the $+q$ charge is now three times as large as before, since there are now three positive charges exerting forces on it."*

Student 2: *"I don't think so. The force from the $+Q$ charge on the left will cancel the force from the $+Q$ charge on the right. The net electric force will be the same as in part A."*

1. Do you agree with either student? Explain.

2. Indicate the direction of the net electric force on the $+q$ charge. Explain.

3. What, if anything, can be said about how the magnitude of the net electric force on the $+q$ charge changes when the two $+Q$ charges are added? Explain.

C. Rank the four cases below according to the magnitude of the net electric force on the $+q$ charge. Explain how you determined your ranking.

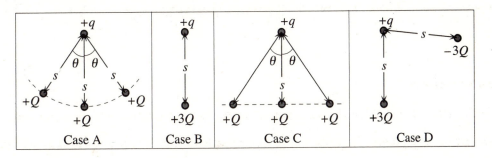

Case A   Case B   Case C   Case D

⇨ Check your ranking with a tutorial instructor before continuing.

Tutorials in Introductory Physics
McDermott, Shaffer, & P.E.G., U. Wash.

©Prentice Hall, Inc.
First Edition, 2002

III. Distributed charge

A. Charge an acrylic rod by rubbing it with wool.

Obtain a small pith ball attached to an insulating thread. Touch the ball to the charged rod and observe the behavior of the ball after it touches the rod.

Is the ball charged after it touches the rod? If so, does the ball have the same sign charge as the rod or the opposite sign charge? Explain how you can tell.

B. Hold the charged rod horizontally. Use a charged pith ball to explore the region around the rod. On the basis of your observations, sketch a *vector* to represent the net electric force on the ball at each of the points marked by an "**×**."

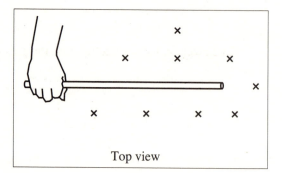

Top view

Is all of the charge on the rod located at a single point? (*e.g.,* Is all the charge at the tip of the rod? At the middle?) Explain how you can tell.

On the basis of the vectors you have drawn, is it appropriate to consider the charged rod as a point charge? Explain.

C. Imagine that two charged rods are held together as shown and a charged pith ball is placed at point *P*.

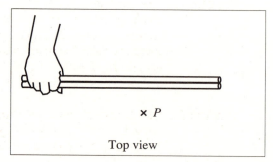

× *P*

Top view

Predict whether the rod farther from point *P* would exert an electric force on the pith ball. Explain.

Check your prediction by placing a charged pith ball at point *P* near two charged rods and then slowly moving one rod away from the other. Describe your observations and discuss with your partners whether your results from this experiment support your prediction.

D. Five short segments (labeled 1–5) of acrylic rod are arranged as shown. All were rubbed with wool and have the same magnitude charge. A charged pith ball is placed in turn at the locations marked by points *A* and *B*.

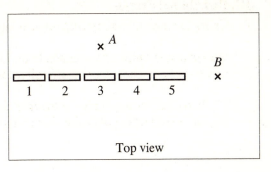

Top view

Indicate the approximate direction of the force on the pith ball at points *A* and *B* due to segment 5 alone.

What is the direction of the *net force* on the pith ball at points *A* and *B*? Explain how you determined your answer.

Does segment 2 exert a force on the pith ball when the pith ball is placed at point *B*? Explain.

E. In case A at right, a point charge $+q$ is a distance *s* from the center of a small ball with charge $+Q$.

In case B the $+q$ charge is a distance *s* from the center of an acrylic rod with a total charge $+Q$.

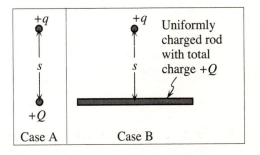

Case A | Case B

Consider the following student dialogue:

Student 1: *"The charged rod and the charged ball have the same charge, +Q, and are the same distance from the point charge, +q. So the force on +q will be the same in both cases."*

Student 2: *"No, in case B there are charges spread all over the rod. The charge directly below the point charge will exert the same force on +q as the ball in case A. The rest of the charge on the rod will make the force in case B bigger."*

Neither student is correct. Discuss with your partners the errors made by each student. Write a correct description of how the forces compare in the space below. Explain.

Tutorials in Introductory Physics
McDermott, Shaffer, & P.E.G., U. Wash.

©Prentice Hall, Inc.
First Edition, 2002

IV. A model for electric charge

A. A small ball with zero net charge is positively charged on one side, and equally negatively charged on the other side. The ball is placed near a positive point charge as shown.

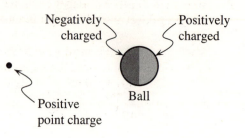

Would the ball be *attracted toward, repelled from,* or *unaffected by* the positive point charge? Explain.

Is your answer consistent with Coulomb's law? Explain.

B. Hang an uncharged metal or metal-covered ball from an insulating string. Then charge a piece of tape as in section I and bring the tape toward the ball.

Describe what you observe.

C. The situation in part A suggests a way to think about the attraction in part B between a charged piece of tape and an uncharged metal ball.

Try to account for the attraction in part B. As part of your answer, draw a sketch of the charge distribution on the tape and ball both before and after they are brought near one another.

I. Area as a vector

A. Hold a small piece of paper (*e.g.*, an index card) flat in front of you. The paper can be thought of as a part of a larger plane surface.

What *single* line could you use to specify the orientation of the plane of the paper (*i.e.*, so that someone else could hold the paper in the same, or in a parallel, plane)?

The line used to orientat of the plane of paper is the line normal to the plane.

B. The area of a flat surface can be represented by a single *vector*, called the area vector $\vec{A}$.

What does the direction of the vector represent?

Normal to the lat surface.

What would you expect the magnitude of the vector to represent?

Area of flat surface.

C. Place a large piece of graph paper flat on the table.

Describe the direction and magnitude of the area vector, $\vec{A}$, for the entire sheet of paper. *All the individual squars appear the same, thus hold a single vector.*
(equal)   *(same direction & magnitude)*

Describe the direction and magnitude of the area vector, $d\vec{A}$, for each of the individual squares that make up the sheet. *Area vector of each individual squars are parallel as vector.*

D. Fold the graph paper twice so that it forms a hollow triangular tube.

Can the entire sheet be represented by a single vector with the characteristics you defined above? If not, what is the minimum number of area vectors required?

Yes, the entire sheet can be represented by a single vector.

E. Form the graph paper into a tube as shown.

Can the orientation of each of the individual squares that make up the sheet of graph paper still be represented by $d\vec{A}$ vectors as in part C above? Explain. *No, orientation of the individual square is not the same for all squares.*

F. What must be true about a surface or a portion of a surface in order to be able to associate a single area vector $\vec{A}$ with that surface?

The area vector of smaller pieces (individual) must equal the same.

II. Electric field

A. In the tutorial *Charge*, you explored the region around a charged rod with a pith ball that had a charge of the same sign as the rod.

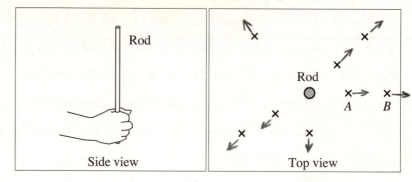

Rod

Side view

Rod

A B

Top view

Sketch *vectors* at each of the marked points to represent the electric force exerted on the ball at that location.

How does the magnitude of the force exerted on the ball at point *A* compare to the magnitude of the force on the ball at point *B*? Point A being closer to the rod has a greater magnitude of force than point B.

B. Suppose that the charge, q_{test}, on the pith ball were halved.

Would the electric force exerted on the ball at each location change? If so, how? If not, explain why not. Yes, as the magnitude of force is decreased.

Would the ratio $\vec{F}/q_{test}$ change? If so, how? If not, explain why not. Yes, as the charge itself has lowered.

C. The quantity $\vec{F}/q_{test}$ evaluated at any point is called the *electric field* $\vec{E}$ at that point.

How does the magnitude of the electric field at point *A* compare to the magnitude of the electric field at point *B*? Explain. Point A would still have a greater force exerted than point B.

D. Sketch *vectors* at each of the marked points to represent the electric field $\vec{E}$ at that location.

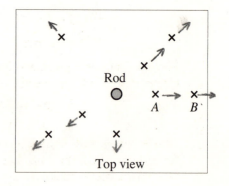

Rod

A B

Top view

Would the magnitude or the direction of the electric field at point *A* change if:

- the charge on the *rod* were increased? Explain. Only the magnitude would increase as greater force.

- the magnitude of the test charge were increased? Explain. Niether magnitude nor direction will change.

- the sign of the test charge were changed? Explain. Direction would change, as a negative attracts.

Tutorials in Introductory Physics
McDermott, Shaffer, & P.E.G., U. Wash.

©Prentice Hall, Inc.
First Edition, 2002

The electric field is typically represented in two ways: by *vectors* or by *electric field lines*. In the vector representation, vectors are drawn at various points to indicate the direction and magnitude of the electric field at those points. In the *field line* representation, straight or curved lines are drawn so that the tangent to each point on the line is along the direction of the electric field at that point. Below, we explore how the field line representation can also reflect the magnitude of the electric field.

E. The diagram at right shows a two-dimensional top view of the electric field lines representing the electric field for a positively charged rod.

You determined previously that the magnitude of the electric field at point *A* was larger than the field at point *B*. What feature of the electric field lines reflects this information about the magnitude of the field?

The position of arrowheads and proximity around charge.

III. Flux

Ask a tutorial instructor for a block of wood with nails through it. The nails represent uniform electric field lines. (The block of wood does not represent anything but serves to hold the nails in place.)

At right is a two-dimensional representation of the same electric field as viewed from the side.

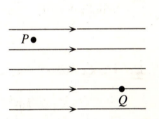

A. Compare the magnitude of the electric field at points *P* and *Q*. Explain your reasoning. *Point P holds a greater magnitude than point Q, being closer to the arrowhead.*

Suppose you were given another block of wood with nails representing a weaker uniform electric field than the one above. How would the two blocks differ? Explain.

The weaker uniform block electric field is less compact.

B. Obtain a wire loop. The loop represents the *boundary* of an imaginary flat surface of area *A*. (In order to allow the nails that represent the field to pass through the surface, you have only been given the boundary of the surface.)

Draw a diagram to show the relative orientation of the loop and the electric field so that the number of field lines that pass through the surface of the loop is:

• the maximum possible.

• the minimum possible.

For a given surface, the *electric flux, Φ_E* , is proportional to the number of field lines through the surface. For a uniform electric field, the maximum electric flux is equal to the product of electric field at the surface and the surface area *(i.e., EA)*. The electric flux is defined to be positive when the electric field $\vec{E}$ has a component in the same direction as the area vector $\vec{A}$ and is negative when the electric field has a component in the direction opposite to the area vector.

C. Sketch vectors $\vec{A}$ and $\vec{E}$ such that the electric flux is:

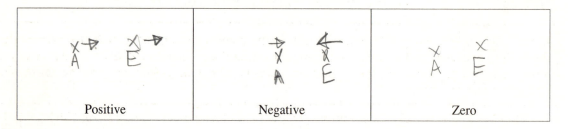

| Positive | Negative | Zero |

D. You will now examine the relationship between the number of field lines through a surface and the angle between $\vec{A}$ and $\vec{E}$.

(You will need a protractor to measure angles.)

1. Place the loop over the nails so that the number of field lines through it is a maximum. Determine the angle between $\vec{A}$ and $\vec{E}$. Record both that angle and the number of field lines that pass through the loop.

2. Rotate the loop until there is one fewer row of nails passing through it. Determine the angle between $\vec{A}$ and $\vec{E}$ and record your measurement. Continue in this way until $\theta = 180°$.

3. On graph paper, plot a graph of *n* versus θ. (Let the number of field lines through the surface be a negative number for angles between 90° and 180°.)

n (# of field lines through surface)	θ (angle between $\vec{A}$ and $\vec{E}$)
100	0
90	25
80	40
70	50
60	60
50	65
40	70
30	75
20	80
10	85
0	90
-10	95
-20	100
-30	105
-40	110

E. When $\vec{E}$ and $\vec{A}$ were parallel, we called the quantity *EA* the electric flux through the surface. For the parallel case, we found that *EA* is proportional to the number of field lines through the surface.

By what trigonometric function of θ must you multiply *EA* so that the product is proportional to the number of field lines through the area for any orientation of the surface?

cosine θ.

Rewrite the quantity described above as a product of just the vectors $\vec{E}$ and $\vec{A}$.

100 cos(θ)

Tutorials in Introductory Physics
McDermott, Shaffer, & P.E.G., U. Wash.

©Prentice Hall, Inc.
First Edition, 2002

I. Electric flux through closed surfaces

In the tutorial *Electric field and flux* and in the homework, we found that the electric flux through a set of imaginary surfaces, $d\vec{A}_i$, each with a uniform electric field, $\vec{E}_i$, can be written as:

$$\Phi_{net} = \vec{E}_1 \bullet d\vec{A}_1 + \vec{E}_2 \bullet d\vec{A}_2 + \vec{E}_3 \bullet d\vec{A}_3 + ...$$

The area vectors at each point on a *closed surface* (*i.e.*, a surface that surrounds a region so that the only way out of the region is through the surface) are chosen by convention to point *out of* the enclosed region. A closed imaginary surface is called a *Gaussian surface*.

In the following questions, a Gaussian cylinder with radius a and length l is placed in various electric fields. The end caps are labeled A and C and the side surface is labeled B. In each case, *base your answer about the net flux only on qualitative arguments about the magnitude of the flux through the end caps and side surface.*

A. The Gaussian cylinder is in a uniform electric field of magnitude E_o aligned with the cylinder axis.

- Find the sign and magnitude of the flux through:

    Surface A: Surface B: Surface C:

- Is the net flux through the Gaussian surface *positive, negative,* or *zero?*

B. The Gaussian cylinder encloses a negative charge. (The field from part A is removed.)

- Find the sign of the flux through:

    Surface A: Surface B: Surface C:

- Is the net flux through the Gaussian surface *positive, negative,* or *zero?*

C. The Gaussian cylinder encloses opposite charges of equal magnitude. (The charges are on the axis of the cylinder and equidistant from the center.)

- Find the sign of the flux through:

    Surface A: Surface B: Surface C:

- Is the net flux through the Gaussian surface *positive, negative,* or *zero?*

D. A positive charge is located above the Gaussian cylinder.

- Find the sign of the flux through:

    Surface A: Surface B: Surface C:

- Can you tell whether the net flux through the Gaussian surface is *positive, negative,* or *zero?* Explain.

⇨ Check your results with a tutorial instructor before continuing.

Tutorials in Introductory Physics
McDermott, Shaffer, & P.E.G., U. Wash.

©Prentice Hall, Inc.
First Edition, 2002

II. Gauss' law

Gauss' law states that the electric flux through a Gaussian surface is directly proportional to the net charge enclosed by the surface ($\Phi_E = q_{enclosed}/\varepsilon_o$).

A. Are your answers to parts A–C of section I consistent with Gauss' law? Explain.

B. In part D of section I, you tried to determine the sign of the flux through the Gaussian cylinder shown.

1. If you have not done so already, use Gauss' law to determine whether the net flux through the Gaussian surface is *positive, negative,* or *zero.* Explain.

2. If $\Phi_A = -10$ Nm^2/C and $\Phi_C = 2$ Nm^2/C, what is Φ_B?

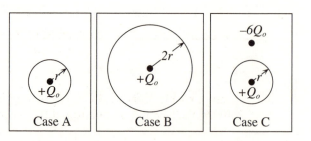

C. Find the net flux through each of the Gaussian surfaces below.

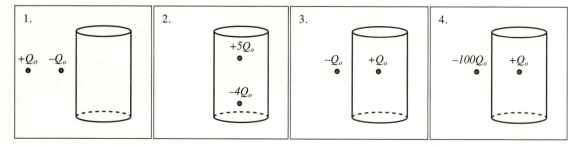

D. The three spherical Gaussian surfaces at right each enclose a charge $+Q_o$. In case C there is another charge $-6Q_o$ outside the surface.

Case A   Case B   Case C

Consider the following conversation:

Student 1: *"Since each Gaussian surface encloses the same charge, the net flux through each must be the same."*

Student 2: *"Gauss' law doesn't apply here. The electric field at the Gaussian surface in case B is weaker than in case A, because the surface is farther from the charge. Since the flux is proportional to the electric field strength, the flux must also be smaller in case B."*

Student 3: *"I was comparing A and C. In C the charge outside changes the field over the whole surface. The areas are the same, so the flux must be different."*

Do you agree with any of the students? Explain.

III. Application of Gauss' law

A. A large sheet has charge density $+\sigma_o$. A cylindrical Gaussian surface encloses a portion of the sheet and extends a distance L_o on either side of the sheet. A_1, A_2, and A_3 are the areas of the ends and curved side, respectively. Only a small portion of the sheet is shown.

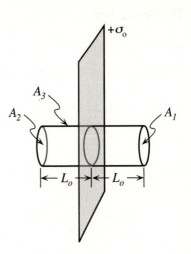

1. On the diagram at right indicate the location of the charge enclosed by the Gaussian cylinder.

    In terms of σ_o and other relevant quantities, what is the net charge enclosed by the Gaussian cylinder?

2. Sketch the electric field lines on both sides of the sheet.

    Does the Gaussian cylinder affect the field lines or the charge distribution? Explain.

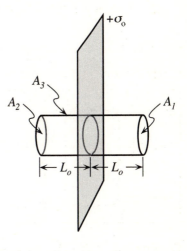

3. Let E_L and E_R represent the magnitude of the electric field on the left and right ends of the Gaussian surface.

    How do the magnitudes of E_L and E_R compare? Explain.

    How do the magnitudes of the areas of the ends of the Gaussian surface compare?

4. Through which of the surfaces (A_1, A_2, A_3) is there a net flux? Explain using a sketch showing the relative orientation of the electric field vector and the area vectors.

    Write an expression for the net electric flux Φ_{net} through the cylinder in terms of the three areas $(A_1, A_2,$ and $A_3)$, E_L, and E_R.

    Use the relationships between the electric fields E_L and E_R and between the areas A_1 and A_2 to simplify your equation for the net flux.

Tutorials in Introductory Physics
McDermott, Shaffer, & P.E.G., U. Wash.

©Prentice Hall, Inc.
First Edition, 2002

5. Gauss' law ($\Phi_E = q_{enclosed}/\varepsilon_o$) relates the net electric flux through a Gaussian surface (which you found in part 4) to the net charge enclosed (which you found in part 1). Use this relationship to find the direction and magnitude of the electric field at the right end of the cylinder in terms of σ_o.

What is the electric field at the left end of the cylinder?

Does the electric field near a large sheet of charge depend on the distance from the sheet? Use your results above to justify your answer.

Is your answer consistent with the electric field lines you sketched in part 2? Explain.

⇨ Check your results with a tutorial instructor before you continue.

B. The Gaussian cylinder below encloses a portion of two identical large sheets. The charge density of the sheet on the left is $+\sigma_o$; the charge density of the sheet on the right is $+2\sigma_o$.

1. Find the net charge enclosed by the Gaussian cylinder in terms of σ_o and any relevant dimensions.

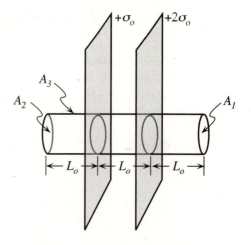

2. Let E_L and E_R be the magnitudes of the electric fields at the left and right end caps of the Gaussian cylinder respectively.

    Is E_L *greater than*, *less than*, or *equal to* E_R? Explain.

3. Find the net flux through the Gaussian cylinder in terms of E_L, E_R, and any relevant dimensions.

4. Use Gauss' law to find the electric field a distance L_o to the right of the rightmost sheet.

Are your results consistent with the results you would obtain using superposition? Explain.

I. Review of work

A. Suppose an object moves under the influence of a force. Sketch arrows showing the relative directions of the force and displacement when the work done by the force is:

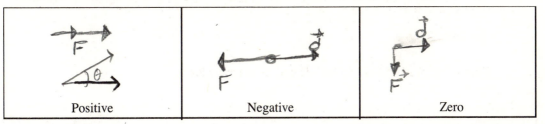

Positive	Negative	Zero

B. An object travels from point A to point B while two constant forces, $\vec{F_1}$ and $\vec{F_2}$, of equal magnitude are exerted on it.

1. Is the total work done on the object by $\vec{F_1}$ *positive, negative,* or *zero?*

    Negative, as force is opposite from the work.

2. Is the total work done on the object by $\vec{F_2}$ *positive, negative,* or *zero?*

    Positive, as force is the same direction as the work.

3. Is the net work done on the object *positive, negative,* or *zero?* Explain.

    Zero, as net force is zero on object.

4. Is the speed of the object at point B *greater than, less than,* or *equal to* the speed of the object at point A? Explain how you can tell.

    Equal, as net force is zero. ($K_B = K_1$)

• Point B

$\vec{F_2}$

$\vec{F_1}$

Point A •

C. An object travels from point A to point B while two constant forces, $\vec{F_3}$ and $\vec{F_4}$, of *unequal* magnitude are exerted on it as shown.

1. Is the total work done on the object by $\vec{F_3}$ *positive, negative,* or *zero?*

    Negative, force is opposite from work.

2. Is the total work done on the object by $\vec{F_4}$ *positive, negative,* or *zero?*

    Positive, force is in same direction as work.

3. Is the net work done on the object *positive, negative,* or *zero?* Explain.

    If $F_4 > F_3$ than positive.
    If $F_3 > F_4$ than negative.

4. Is the speed of the object at point B *greater than, less than,* or *equal to* the speed of the object at point A? Explain how you can tell.

    If $F_4 > F_3$ than point B would be greater than point A.

• Point B

$\vec{F_4}$

$\vec{F_3}$

Point A •

Tutorials in Introductory Physics
McDermott, Shaffer, & P.E.G., U. Wash.

©Prentice Hall, Inc.
First Edition, 2002

D. State the work-energy theorem in your own words. Are your answers in part B consistent with this theorem? Explain.

"Network done is change in kinetic energy."

Yes, as $K = K_i$, $V_A = V_B$

Are your answers in part C consistent with the work-energy theorem? Explain.

Yes, as $K_i > K_i$ $V_B > V_A$

II. Work and electric fields

The diagram at right shows a top view of a positively charged rod. Points W, X, Y, and Z lie in a plane near the center of the rod. Points W and Y are equidistant from the rod, as are points X and Z.

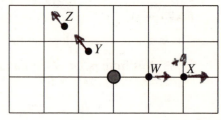

A. Draw electric field vectors at points W, X, Y, and Z.

B. A particle with charge $+q_o$ travels along a straight line path from point W to point X.

Is the work done *by the electric field* on the particle *positive, negative,* or *zero*? Explain using a sketch that shows the electric force on the particle and the displacement of the particle.

Positive, because the charge is moving in the same direction of electric field.

Compare the work done by the electric field when the particle travels from point W to point X to that done when the particle travels from point X to point W. $W_{X \to W} = -W_{W \to X}$

It is equal but opposite sign shows one time work is done and one is gain in energy.

C. The particle travels from point X to point Z along the circular arc shown.

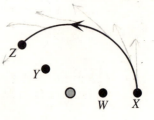

1. Is the work done *by the electric field* on the particle *positive, negative,* or *zero*? Explain. (*Hint:* Sketch the direction of the force on the particle and the direction of the displacement for several short intervals during the motion.)

Zero, $W_{X \to w} + W_{w \to rod} = -(W_{rod \to y} + W_{y \to z})$

work $x \to z$ is same and it is independent of both.

Tutorials in Introductory Physics
McDermott, Shaffer, & P.E.G., U. Wash.

©Prentice Hall, Inc.
First Edition, 2002

2. Compare the work done by the electric field when the particle travels from point W to point X to that done when the particle travels from point W to point Z along the path shown. Explain.

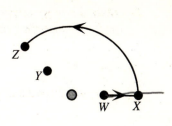

$$W_{W \to X} = W_{W \to Z}$$
since
$$W_{X \to Z} = 0$$

D. Suppose the particle travels from point W to point Y along the path $WXZY$ as shown.

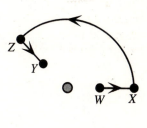

1. Compare the work done by the electric field when the particle travels from point W to point X to that done when the particle travels from point Z to point Y. Explain.

As $X \to W$ is same for $Z \to y$

therefore $W_{W \to X} = -W_{Z \to y}$

What is the total work done on the particle by the electric field as it moves along the path $WXZY$?   total work

$$W = 0 \qquad due\ to\ \ W_{W \to X} + W_{Z \to y} = 0$$

2. Suppose the particle travels from W to Y along the arc shown. Is the work done on the particle by the electric field *positive*, *negative*, or *zero*? Explain using force and displacement vectors.

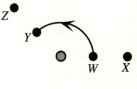

Zero.
$$W = F_o \sin 0°$$
$$W = F_o \sin 0°$$
$$W = 0$$

3. Suppose the particle travels along the straight path WY. Is the work done on the particle by the electric field *positive*, *negative*, or *zero*? Explain using force and displacement vectors. (*Hint:* Compare the work done along the first half of the path to the work done along the second half.)   zero $W_{W \to y} = 0$

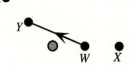

work is both independent for a conservative force.

Tutorials in Introductory Physics
McDermott, Shaffer, & P.E.G., U. Wash.

©Prentice Hall, Inc.
First Edition, 2002

E. Compare the work done as the particle travels from point *W* to point *Y* along the three different paths in part D.

W → X, positive

Z → y, negative

zero

It is often said that the work done by a static electric field is *path independent*. Explain how your results in part D are consistent with this statement.

Each path presents different charges being independent.

III. Electric potential difference

A. Suppose the charge of the particle in section II is increased from $+q_o$ to $+1.7q_o$.

1. Is the work done by the electric field as the particle travels from *W* to *X greater than, less than,* or *equal to* the work done by the electric field on the original particle? Explain.

Greater than, due to the 1.7 increase of charge.

2. How is the quantity *the work divided by the charge* affected by this change?

Know 1.7×fq0 & the work 1.7 must equal to 1.7 wq0

The *electric potential difference* ΔV_{WX} between two points *W* and *X* is defined to be:

$$\Delta V_{WX} = -\frac{W_{elec}}{q}$$

where W_{elec} is the work done by the field as a charge *q* travels from point *W* to point *X*.

3. Does this quantity depend on the *magnitude* of the charge of the particle that is used to measure it? Explain.

It does not as the measuring particles hold the same relation.

4. Does this quantity depend on the *sign* of the charge of the particle that is used to measure it? Explain.

No, as the ratio will not change.

Tutorials in Introductory Physics
McDermott, Shaffer, & P.E.G., U. Wash.

©Prentice Hall, Inc.
First Edition, 2002

B. Shown at right are four points near a positively charged rod. Points *W* and *Y* are equidistant from the rod, as are points *X* and *Z*. A charged particle with mass $m_o = 3 \times 10^{-8}$ kg is released from rest at point *W* and later is observed to pass point *X*.

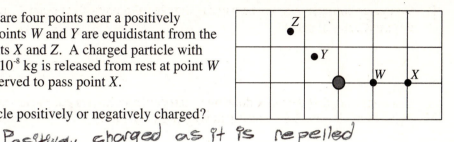

1. Is the particle positively or negatively charged? Explain.

Positively charged as it is repelled by the rod.

2. Suppose that the magnitude of the charge on the particle is 2×10^{-6} C and that the speed of the particle is 40 m/s as it passes point *X*.

a. Find the change in kinetic energy of the particle as it travels from point *W* to point *X*.

$\frac{1}{2} \times (3 \times 10^{-8}) \times 40^2 - \frac{1}{2} \times m_o \times 0^2$

$\Delta KE = 24 \times 10^{-6}$ J

b. Find the work done on the particle by the electric field between point *W* and point *X*. (*Hint:* See part D of section I.)

$W_{ele.} = \Delta KE$

$W_{ele.} = 24 \times 10^{-6}$ J

c. Find the electric potential difference between point *W* and point *X*.

$V_X - V_W = \dfrac{-24 \times 10^{-6}}{2 \times 10^{-6}}$ volts

$V_W - V_X = 12 v$

d. If the same particle were released from point *Y*, would its speed as it passes point *Z* be *greater than, less than,* or *equal to* 40 m/s? Explain.

The potential difference is the same as work and is equal to 40 m/s, thus

3. Suppose that a second particle with the same mass as the first but nine times the charge (*i.e.*, 18×10^{-6} C) were released from rest at point W.

a. Would the electric potential difference between points *W* and *X* change? If so, how, if not, why not?

No, as the potential around the rod remains same.

b. Would the speed of the second particle as it passes point *X* be *greater than, less than,* or *equal to* the speed of the first particle as it passed point *X?* Explain.

$V = \sqrt{(2 \cap S)}$ so the speed is greater.

4. A particle with mass $m_o = 3 \times 10^{-8}$ kg is launched toward the rod from point Z and turns around at point Y.

    a. If the particle has charge $q_o = 2 \times 10^{-6}$ C, with what speed should it be launched? Explain.

    $$V_P = \sqrt{\frac{2}{m} q (V_1 - V_z)}$$

    $$(V_y - V_z)$$

    $$(2 \times 10^{-6})(3 \times 10^{-8})$$

    40 m/s

    b. If instead the particle has charge $9q_o$ (*i.e.*, 18×10^{-6} C) with what speed should it be launched? Explain.

    3x faster   120 m/s

Tutorials in Introductory Physics
McDermott, Shaffer, & P.E.G., U. Wash.

©Prentice Hall, Inc.
First Edition, 2002

I. The electric field near conducting plates

A. A small portion near the center of a large thin conducting *plate* is shown magnified at right. The portion shown has a net charge Q_1 and each side has an area A_1.

Write an expression for the charge density on each side of the conducting plate.

Side view of thin
charged plate

B. Use the principle of superposition to determine the electric field inside the conductor (if you have not done so already).

Is your answer consistent with your knowledge of the electric field inside a conductor? Explain.

C. Use the principle of superposition to determine the electric field on each side of the plate.

Does the charge on the *right* surface contribute to the electric field to the *left* of the plate (even though metal separates the two regions)? Explain.

Tutorials in Introductory Physics
McDermott, Shaffer, & P.E.G., U. Wash.

©Prentice Hall, Inc.
First Edition, 2002

D. Consider instead a portion near the center of a large *sheet* of charge. Like the plate in part A, the portion of the sheet has a net charge Q_1 and area A_1.

How does the charge density σ' on this sheet compare to the charge density on each side of the plate above? Explain.

How does the electric field on one side of the *sheet of charge* compare to the electric field on the same side of the *charged plate?* Explain.

E. A second plate with the same magnitude charge as the first, but opposite sign, is now held near the first. The plates are large enough and close enough together that fringing effects near the edges can be ignored.

The diagrams below show various distributions of charge on the two plates. Decide which arrangement is physically correct. Explain.

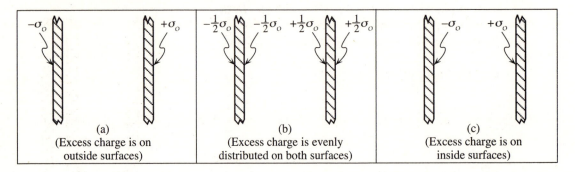

(a)
(Excess charge is on outside surfaces)

(b)
(Excess charge is evenly distributed on both surfaces)

(c)
(Excess charge is on inside surfaces)

Tutorials in Introductory Physics
McDermott, Shaffer, & P.E.G., U. Wash.

©Prentice Hall, Inc.
First Edition, 2002

II. Parallel plates and capacitance

Two very large thin conducting plates are a distance D apart. The surface area of the face of each plate is A_o. A side view of a small portion near the center of the plates is shown.

A. The inner surface of one plate has a uniform charge density of $+\sigma_o$; the other, $-\sigma_o$. The charge density on the outer surface of each plate is zero.

1. At each labeled point, draw vectors to represent the electric field at that point due to each charged plate.

2. Write expressions for the following quantities in terms of the given variables:

   • the electric field at points *1, 2, 3,* and *4*

$$\text{Point 1: } E_1 = \frac{\sigma}{2\varepsilon_0} - \frac{\sigma}{2\varepsilon_0} = 0 \qquad \text{Point 2: } E_2 = \frac{\sigma}{2\varepsilon_0} + \frac{\sigma}{2\varepsilon_0} = \frac{\sigma}{\varepsilon_0}$$

$$\text{\&}$$

$$\text{Point 4} \qquad\qquad\qquad\qquad\qquad \text{\&}$$

$$\qquad\qquad\qquad\qquad\qquad\qquad\qquad \text{Point 3}$$

   • the potential difference between the plates

$$V = ED = \frac{\sigma}{\varepsilon_0}(D) = \frac{QD}{A\varepsilon_0}$$

$$\Delta V = E\Delta x$$

3. The right plate is moved to the left as shown. Both plates are kept insulated. Describe how each of the following quantities will change (if at all). Explain.

   • the charge density on each plate

$$\text{Unchanged, Just flipped still same.}$$

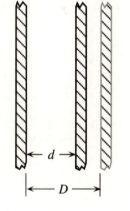

   • the electric field both outside and between the plates

$$\text{Unchanged, Just flipped still same.}$$

   • the potential difference between the plates

$$\text{Decreases}$$

$$\Delta V = E\Delta x$$

4. Write expressions for the following quantities in terms of σ_o and d (the new distance between the plates).

• the magnitude of the electric field between the plates

• the potential difference between the plates

NA

5. Find $\dfrac{Q}{\Delta V}$ (the ratio of the net charge on one plate to the potential difference between the plates).

$$\frac{Q}{\Delta V} = \frac{\sigma_o A_o}{\frac{\sigma_o}{\mathcal{E}_o} d} = \frac{A_o \mathcal{E}_o}{d} = C$$

How, if at all, would this ratio change if the charge densities on the plates were $+2\sigma_o$ and $-2\sigma_o$?   *Unchanged, the charge densities cancel each other out.*

⇨ Check your results for part A with a tutorial instructor before you continue.

B. Suppose the plates are discharged, then held a distance D apart and connected to a battery. (Ignore the fringing fields near the plate edges.)

1. Write expressions for the following quantities in terms of the given variables. Explain your reasoning in each case.

• the potential difference ΔV between the plates

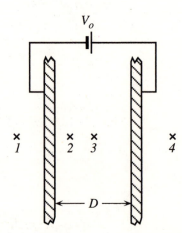

• the electric field at points *1, 2, 3,* and *4*

• the charge density on each plate

Tutorials in Introductory Physics
McDermott, Shaffer, & P.E.G., U. Wash.

©Prentice Hall, Inc.
First Edition, 2002

2. The right plate is moved to the left. Describe how each of the following quantities changes (if at all). Explain.

    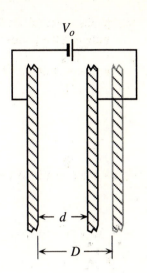

    • the potential difference ΔV between the plates

    • the electric field both outside and between the plates

    • the charge density on each plate

3. Write expressions for the following quantities in terms of V_o and d (the new distance between the plates).

    • the magnitude of the electric field between the plates

    • the charge density on each plate

4. Find $\dfrac{Q}{\Delta V}$ (the ratio of the net charge on one plate to the potential difference between the plates).

    How, if at all, would this ratio change if the voltage of the battery was $2V_o$?

⇨ Check your results for part B with a tutorial instructor before you continue.

C. Compare the ratio $\dfrac{Q}{\Delta V}$ that you calculated for two insulated plates (part A) to the same ratio for two plates connected to a battery (part B).

   1. Does the ratio $\dfrac{Q}{\Delta V}$ depend on whether or not the plates are connected to a battery?

   2. Does the ratio $\dfrac{Q}{\Delta V}$ depend on the distance between the plates?

The potential difference ΔV between two isolated conductors depends on their net charges and their physical arrangement. If the conductors have charge $+Q$ and $-Q$, the ratio $\dfrac{Q}{\Delta V}$ is called the *capacitance (C)* of the particular arrangement of conductors.

D. For the following cases, state whether each of the quantities q, σ, E, ΔV, and C changes or remains fixed:

   1. two insulated conducting plates are moved farther apart

   2. two conducting plates connected to a battery are moved farther apart

Tutorials in Introductory Physics
McDermott, Shaffer, & P.E.G., U. Wash.

©Prentice Hall, Inc.
First Edition, 2002

In this tutorial, we construct a model for electric current that we can use to predict and explain the behavior of simple electric circuits.

I. Complete circuits

A. Obtain a battery, a light bulb, and a single piece of wire. Connect these in a variety of ways. Sketch each arrangement below.

Arrangements that *do* light the bulb	Arrangements that *do not* light the bulb
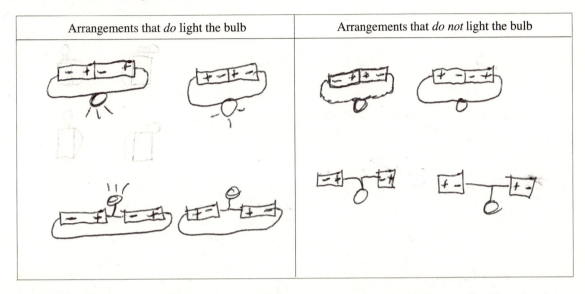	

You should have found at least four different arrangements that light the bulb. How are these arrangements similar? How do they differ from arrangements in which the bulb does not light? The different polarities are always connect and a circuit has no gap. Unlike those that don't light opposite in every way.

State the requirements that must be met in order for the bulb to light.

• Different sign ends connets to each other

• Circuit must be a complete loop that is closed.

B. A student has briefly connected a wire across the terminals of a battery until the wire feels warm. The student finds that the wire seems to be equally warm at points *1, 2,* and *3*.

Based on this observation, what might you conclude is happening in the wire at one place compared to another?

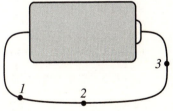

Due to the voltage remaining the same through wire all points are also heating up equally.

Tutorials in Introductory Physics
McDermott, Shaffer, & P.E.G., U. Wash.

©Prentice Hall, Inc.
First Edition, 2002

C. Light a bulb using a battery and a single wire. Observe and record the behavior
(*i.e.,* brightness) of the bulb when objects made out of various materials are inserted into the
circuit. (Try materials such as paper, coins, pencil lead, eraser, your finger, *etc.*)

What is similar about most of the objects that let the bulb light?   *N/A*

In theroy more metallic objects
completed the loop

D. Carefully examine a bulb. Two wires extend from the filament of the bulb
into the base. You probably cannot see into the base, however, you should be
able to make a good guess as to where the wires are attached. Describe where
the wires attach. Explain based on your observations in parts A–C.

One Junction is to negative the other
positive of a current to the bulb
being a closed loop.

On the basis of the observations that we have made, we will make the following assumptions:

1. A flow exists in a complete circuit from one terminal of the battery, through the rest of
the circuit, back to the other terminal of the battery, through the battery and back around
the circuit. We will call this flow *electric current*.

2. For identical bulbs, bulb brightness can be used as an indicator of the amount of current
through that bulb: the brighter the bulb, the greater the current.

Starting with these assumptions, we will develop a model that we can use to account for the

behavior of simple circuits. The construction of a scientific model is a step-by-step process in

which we specify only the minimum number of attributes that are needed to account for the

phenomena under consideration.

II. Bulbs in series

Set up a two-bulb circuit with identical bulbs connected one after the other as shown. Bulbs connected in this way are said to be connected in *series*.

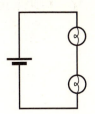

A. Compare the brightness of the two bulbs with each other. (Pay attention only to large differences in brightness. You may notice minor differences if two "identical" bulbs are, in fact, not quite identical.)

Use the assumptions we have made in developing our model for electric current to answer the following questions:

1. Is current "used up" in the first bulb, or is the current the same through both bulbs?

   The current is the same throughout

2. Do you think that switching the order of the bulbs might make a difference? Check your answer. *No*

3. On the basis of your observations *alone,* can you tell the direction of the flow through the circuit? *No*

B. Compare the brightness of each of the bulbs in the two-bulb series circuit with that of a bulb in a single-bulb circuit.

Use the assumptions we have made in developing our model for electric current to answer the following questions:

1. How does the current through a bulb in a single-bulb circuit compare with the current through the same bulb when it is connected in series with a second bulb? Explain.

   A single bulb has more current flow, thus being brighter than with two bulbs.

2. What does your answer to question 1 imply about how the current through the *battery* in a single-bulb circuit compares to the current through the *battery* in a two-bulb series circuit? Explain.

   In one battery, current is twice as large. Due to Ohm's law.

Tutorials in Introductory Physics
McDermott, Shaffer, & P.E.G., U. Wash.

©Prentice Hall, Inc.
First Edition, 2002

C. We may think of a bulb as presenting an obstacle, or *resistance,* to the current in the circuit.

    1. Thinking of the bulb in this way, would adding more bulbs in series cause the total obstacle to the flow, or *total resistance,* to increase, decrease, or stay the same as before?

$$\text{Increase, as} \quad R\text{series} = \sum_i R_i$$
$$\text{Total}$$

    2. Formulate a rule for predicting how the current through the battery would change (*i.e.,* whether it would *increase, decrease,* or *remain the same*) if the number of bulbs connected in series were increased or decreased.

$$I = \frac{V}{R} \text{, as V is same, then twice as} \quad \boxed{\text{Ohm's Laws}}$$
$$\text{much R means half as much I.}$$

III. Bulbs in parallel

Set up a two-bulb circuit with identical bulbs so that their terminals are connected together as shown. Bulbs connected together in this way are said to be connected in *parallel.*

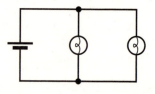

A. Compare the brightness of the bulbs in this circuit.

    1. What can you conclude from your observation about the amount of current through each bulb?

       A equal current in each bulb

    2. Describe the current in the entire circuit. Base your answer on your observations. In particular, how does the current through the battery seem to divide and recombine at the junctions of the two parallel branches?

      At a junction, current divides as

$$I = \sum_i I_i \quad \text{where } i \text{ is number of}$$

      junction −1.

Tutorials in Introductory Physics
McDermott, Shaffer, & P.E.G., U. Wash.

©Prentice Hall, Inc.
First Edition, 2002

B. Is the brightness of each bulb in the two-bulb parallel circuit *greater than, less than,* or *equal to* that of a bulb in a single-bulb circuit?

Equal

How does the amount of current through a *battery* connected to a single bulb compare to the current through a *battery* connected to a two-bulb parallel circuit? Explain based on your observations.

Through battery current is higher.

C. Formulate a rule for predicting how the current through the battery would change (*i.e.,* whether it would *increase, decrease,* or *remain the same*) if the number of bulbs connected in parallel were increased or decreased. Base your answer on your observation of the behavior of the two-bulb parallel circuit and the model for current.

The current is divided equally to each bulb, thus a stronger or weaker would affect each bulb the same.

What can you infer about the total resistance of a circuit as the number of parallel branches is increased or decreased?

The resistance becomes smaller.

D. Does the amount of current through a battery seem to depend on the number of bulbs in the circuit and how they are connected?

It ~~does~~ is dependent, as total resistance is less giving to a higher current.

E. Unscrew one of the bulbs in the two-bulb parallel circuit. Does this change significantly affect the current through the branch that contains the other bulb?

No significant change to current.

A characteristic of an *ideal* battery is that the branches connected directly across it are independent of one another.

IV. Limitations: The need to extend the model

A. The circuit at right contains three identical bulbs and an ideal battery. Assume that the resistance of the switch, when closed, is negligible. Use the model we have developed to:

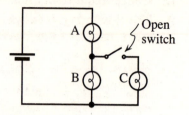

- predict the relative brightness of the bulbs in the circuit with the switch closed. Explain.

Bulb A will be twice as bright as B or C. Due to the current being divided equally amongst B and C.

- predict how the brightness of bulb A changes when the switch is opened. Explain.

Don't change because the current through A is the same.

B. Show that a simple application of the model for current that we have developed thus far is inadequate for determining how the brightness of bulb B changes when the switch is opened.

With the switch open, the current will not divid making B brighter. In theory it would be the same as A.

I. Current and resistance

A. The circuits at right contain identical batteries, bulbs, and unknown identical elements labeled X.

How do the bulbs compare in brightness? Explain. *The same as the current flowing through the bulbs is giving R in series.*

In each circuit, how does the current through the bulb compare to the current through element X? Explain. *current is the same as voltage*

$I = \frac{V}{R}$ *over higher ristance will increase showing the ratio will be the same.*

B. The circuits at right contain identical batteries and bulbs. The boxes labeled X and Y represent different unknown elements. (Assume there are no batteries in either box.)

It is observed that the bulb on the left is brighter than the bulb on the right.

1. Based on this observation, how does the resistance of element X compare to that of element Y? Explain. *Element Y has greater resistance as $P = I^2$ $R = V^2/R$.*

2. In each circuit, how does the current through the bulb compare to the current through the unknown element? *The current in X is greater than Y as the value $P = VI$ is greater.*

3. In each circuit, how does the current through the bulb compare to the current through the battery? *The current is the same.*

C. Predict the relative brightness of bulbs B_1, B_2, and B_3 in the circuits shown. (A dashed box has been drawn around the network of circuit elements that is in series with each of these bulbs.)

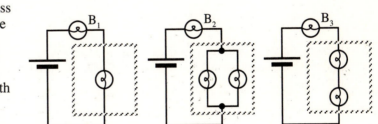

What does your prediction imply about the relative currents through the *batteries?* Explain.

$B_1 > B_3 > B_2$ *as B_2 is split*

Have a tutorial instructor show you these circuits so that you can check your answers. Resolve any conflicts between your answers and your observations.

II. Potential difference

For the remaining circuits in this tutorial use the battery holder with two batteries connected in series. The two-battery combination will be treated as a single circuit element.

A. Set up the circuit with a single bulb and the battery combination as shown. Connect each probe of the voltmeter to a different terminal of the battery holder to measure the potential difference across the battery. Make a similar potential difference measurement across the bulb.

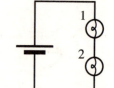

How does the potential difference across the bulb compare to the potential difference across the battery?

A 0.05 difference as for wire resistance.

V_{Bat}	V_{Bulb}
1.80	1.75

B. Set up the circuit containing two bulbs in series as shown.

Rank from largest to smallest the currents through bulb 1, bulb 2, and the bulb in the single-bulb circuit from part A ($i_{Bulb\,1}$, $i_{Bulb\,2}$, i_{single}). Explain.

$i_{single} > i_{B_1} = i_{B_2}$

more resistance

Measure the potential difference across each element in the circuit.

V_{Bat}	$V_{Bulb\,1}$	$V_{Bulb\,2}$
2.17	1.08	1.06

1. How does the potential difference across the battery in this circuit compare to the potential difference across the battery in the single-bulb circuit? (See part A.)

Higher V   2.17 > 1.80

2. Rank the potential differences across bulb 1, bulb 2, and the bulb in the single-bulb circuit from part A.

$V_{B_2} = V_{B_1} < V_{B\,single}$

3. How does the potential difference ranking compare to the brightness ranking of the bulbs?

The same order of voltage ranking.

C. Predict what the voltmeter would read if it were connected to measure the potential difference across the network of bulb 1 and bulb 2 together. Explain.

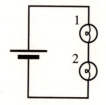

The same as resistance wouldn't change.

Test your prediction.

$V_{AB} = 2.01$ vs. 2.17

How does the potential difference across the network of bulbs compare to the potential difference across the battery?

A 0.16 difference.

Tutorials in Introductory Physics
McDermott, Shaffer, & P.E.G., U. Wash.

©Prentice Hall, Inc.
First Edition, 2002

D. Set up the circuit with two bulbs in parallel as shown.

Rank the currents through bulb 1, bulb 2, and the bulb in the single-bulb circuit from part A. Explain.

$B_1 = B_2 = B$ single being parallel.

How does the current through bulb 1 compare to the current through the battery? Explain.

½ as much due to being ~~possette~~ parallel.

Measure the potential difference across each circuit element.

V_{Bat}	$V_{Bulb\ 1}$	$V_{Bulb\ 2}$
1.74	1.74	1.74

1. How does the potential difference across the battery in this circuit compare to the potential difference across the battery in the single-bulb circuit? 1.74 vs. 1.80

   A 0.06 difference

2. Rank the potential difference across bulb 1, bulb 2, and the bulb in the single-bulb circuit from part A. Entirely equal

3. How does the ranking by potential difference compare to the ranking by brightness?

   The same.

E. Answer the following questions based on the measurements you have made so far.

1. Does the *current through the battery* depend on the circuit to which it is connected? Explain. Yes, as high resistance in circuit is less current through battery.

2. Does the *potential difference across the battery* depend on the circuit to which it is connected? Explain. Yes, as ohm Law is V=RI.

III. Extending the model

Our model for electric circuits includes the idea that, for identical bulbs, the brightness of a bulb is an indicator of the current through the bulb. Based on our observations in this tutorial, we can extend the model to include the idea that, for circuits containing identical bulbs, the brightness of a bulb is also an indicator of the potential difference across the bulb.

A. Set up the circuit with three bulbs as shown and observe their brightness.

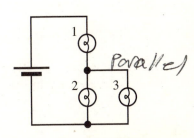

Parallel

Before making the voltmeter measurements, predict the ranking of the potential difference across the battery and each bulb (V_{Bat}, $V_{Bulb\ 1}$, $V_{Bulb\ 2}$, and $V_{Bulb\ 3}$). Explain your prediction.

$V_{Bat} > V_{B_1} > V_{B_2} = V_{B_3}$

Tutorials in Introductory Physics
McDermott, Shaffer, & P.E.G., U. Wash.

©Prentice Hall, Inc.
First Edition, 2002

Measure the potential difference across each element in the circuit. If your measurements are not consistent with your ranking above, resolve the inconsistencies.

V_{Bat}	$V_{Bulb\,1}$	$V_{Bulb\,2}$	$V_{Bulb\,3}$
2.0	1.60	0.022	0.025

B. Before setting up the circuit shown at right:

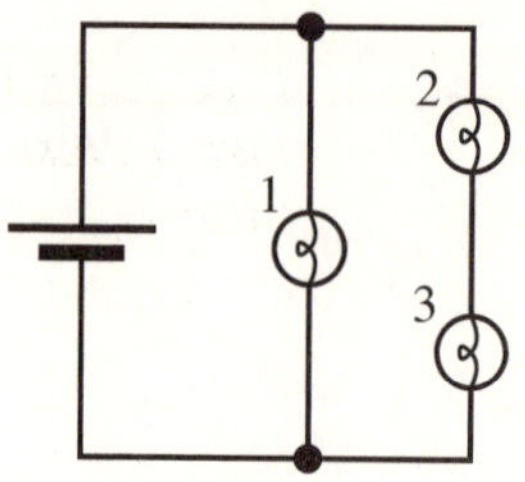

• Predict the ranking of the currents through the battery and each bulb (i_{Bat}, $i_{Bulb\,1}$, $i_{Bulb\,2}$, and $i_{Bulb\,3}$). Explain.

$$i_{Bat} > i_{B_1} > i_{B_2} = i_{B_3}$$

2 &3 are parallel

• Predict the voltmeter measurements across each of the elements in the circuit shown. Explain.

$$V_{Bat} = V_1 > V_2 = V_3$$

Prediction:

V_{Bat}	$V_{Bulb\,1}$	$V_{Bulb\,2}$	$V_{Bulb\,3}$
1.52	1.42	0.020	0.068

Set up the circuit and check your predictions. If your observations and measurements are not consistent with your predictions, resolve the inconsistencies.

Measurement:

V_{Bat}	$V_{Bulb\,1}$	$V_{Bulb\,2}$	$V_{Bulb\,3}$
1.62	1.58	0.79	0.78

C. Both circuits at right have more than one path for the current. Sketch all possible current loops on the diagrams. (A "current loop" is *a single path* of conductors that connects one side of the battery to the other.)

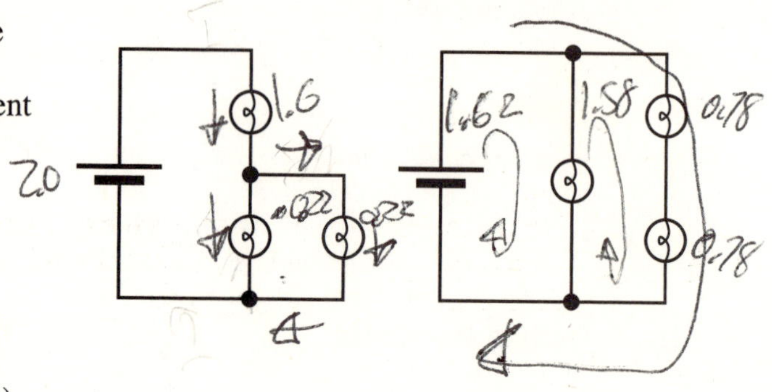

For each of the current loops you have drawn, calculate the sum of the potential differences across the bulbs in that loop. (Use the measurements you made above.)

<u>left</u>
2.0 = 1.6 + 0.22
= 1.82

How do the sums of the potential differences across the bulbs in each loop compare to the potential difference across the battery?

The same.

$$V_B = 1.62$$
$$= V_{B1b1}$$
$$V_{ub2} + V_{b3}$$

⇨ Check your answer with a tutorial instructor.

Tutorials in Introductory Physics
McDermott, Shaffer, & P.E.G., U. Wash.

©Prentice Hall, Inc.
First Edition, 2002

Throughout this tutorial, when you are asked to *predict* the behavior of a circuit, do so *before* setting up the circuit.

I. Simple RC circuits

A. A capacitor is connected to a battery, bulb, and switch as shown. Assume that the switch has been closed for an *extended period of time*.

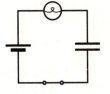

1. *Predict* whether the brightness of the bulb is *the same as, greater than,* or *less than* the brightness of a single bulb connected to a battery. Explain. Less than, as the capacitor plates will go on increasing. So the current goes on decreasing.

2. *Predict* how the potential difference across the battery compares to the potential difference across the capacitor plates and to the potential difference across the bulb. Explain. P.d across the battery is maintained while P.d across capacitor plates is increasing. P.d across the bulb is decreasing.

3. Briefly describe the distribution of charge, if any, on the capacitor plates. The plate attached to the positive terminal get positively charged, and the other plate gets negatively charged.

   Recall the relationship between the charge on a capacitor and the potential difference across the capacitor. Use this relationship to describe how you could use a voltmeter to determine the charge on a capacitor. As capacitance is constant for the capacitor, voltmeter can measure the charge on the capacitor.

4. Obtain the circuit and a voltmeter. Check your predictions for parts 1 and 2. On par with predictions.

B. Remove the capacitor and the bulb from the circuit.

Tutorials in Introductory Physics
McDermott, Shaffer, & P.E.G., U. Wash.

©Prentice Hall, Inc.
First Edition, 2002

1. *Predict* the potential difference across the bulb and the potential difference across the capacitor while these elements are disconnected from the circuit and from each other. Explain.

   The capacitor will become fully charged with equal voltage as the battery. Due to the bulb just as being a resistance.

   Check your prediction.

   on par with prediction.

2. *Predict* whether the potential difference across the capacitor will *increase, decrease,* or *remain the same* if a wire is connected from "ground" to one or the other of the terminals of the capacitor. Explain your reasoning.

   P.d will remain the same, as it maintains the voltage due to it connected from "ground."

   Check your prediction. (You can use a wire with clip leads connected to a metal table leg as a "ground.")

   On par with prediction

3. Devise and carry out a method to reduce the potential difference across the capacitor to zero. (This is sometimes called *discharging* the capacitor.)

   Connect a resistance with a capacitor and voltage will drop to zero.

4. The capacitor in part A is said to be *charged* by the battery.

   Does the capacitor have a *net* charge after being connected to the battery?

   Yes, as the capacitor will charge and maintain a voltage.

   In light of your answer above, what is meant by *the charge* on a capacitor?

   Giving electrical energy to store onto the plates of a capacitor.

Tutorials in Introductory Physics
McDermott, Shaffer, & P.E.G., U. Wash.

©Prentice Hall, Inc.
First Edition, 2002

II. Charging and discharging capacitors

A. Suppose an uncharged capacitor is connected in series with a battery and bulb as shown.

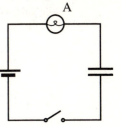

1. *Predict* the behavior of the bulb when the switch is closed. Explain. It will glow and become dimmer until it stops glowing with no loop.

   Set up the circuit and check your prediction. If your prediction is in conflict with your observation, how can you account for your observation?

   As $I = I_0 * e^{(-t/\tau)}$ to review the series.

2. *Without using a voltmeter*, determine the potential difference across the capacitor at the following times:

   • *just after* the switch is closed. Explain how you can tell. (*Hint:* Compare the brightness of the bulb to the brightness of a bulb connected to a battery in a single-bulb circuit without a capacitor.) $V_c = V_s (1 - e^{-t/RC})$

   So than $t \approx 0$ also $V_c = 0$.

   • *a long time after* the switch is closed. Explain how you can tell.

   As $t = \infty$ then $V_c = V_s$;

   Use a voltmeter to check your predictions. (*Hint:* Be sure to discharge the capacitor completely after each observation.) Same as predictions.

B. Suppose that instead of connecting the uncharged capacitor to the single bulb A, you connected it to the two-bulb circuit shown at right.

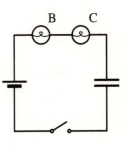

1. *Predict* how the *initial brightness* of bulb B compares to the initial brightness of bulb C. Explain. Initial brightness is the same as voltage drop is the same.

2. *Predict* how the *initial brightness* of bulb B compares to the initial brightness of bulb A above. Explain. The brightness of B will half than A as greater resistance.

   Discharge the capacitor and then set up the circuit with the uncharged capacitor and check your predictions. If your prediction is in conflict with your observation, how can you account for your observation? No conflict with predictions.

Tutorials in Introductory Physics
McDermott, Shaffer, & P.E.G., U. Wash.

©Prentice Hall, Inc.
First Edition, 2002

3. *Predict* how the final charge on the capacitor compares to the final charge on the capacitor from part A.

> Final charge in both cases is equal since both voltage equal their batteries.

Use a voltmeter to check your prediction.

C. Suppose that the bulbs were connected in parallel, rather than in series.

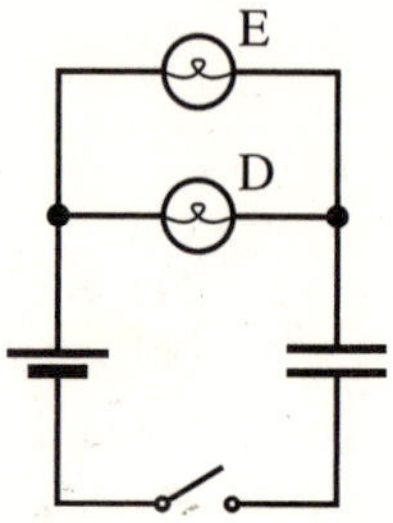

1. *Predict* how the *initial brightness* of bulb D compares to the initial brightness of bulb E. Explain. Both bulbs will glow the same being parallel.

2. *Predict* how the *initial brightness* of bulb D compares to the initial brightness of bulbs A, B, and C above. Explain.

> D=A
> D7B ⎫ Potential drop are half
> D7C ⎭ compared to bulb D.

3. *Predict* how the final charge on the capacitor compares to the final charge on the capacitor from part A. Explain. The same as the capacitor is fully charged as no current will be flowing.

Set up the circuit and check your predictions. If your prediction is in conflict with your observation, how can you account for your observation?

> No noteable conflict with predictions

D. After completing the experiments above, two students make the following comments:

Student 1: *"The capacitor with two bulbs in series got charged up a lot more than the capacitor with two bulbs connected in parallel because the series circuit charged the capacitor for a longer period of time."*

Student 2: *"I disagree, the bulbs in the parallel circuit were brighter so this capacitor gained more charge."*

Do you agree with student 1, student 2, or neither? Explain your reasoning.

> Neither, as both cases the voltage   Q=CV
> across the capacitor is same, being
> the charge should be the same as well.

E. Suppose that a different capacitor of smaller capacitance were connected to the battery and a single bulb in series.

1. *Predict* how the initial potential difference across the bulb compares to the initial potential difference across the bulb in part A.

    Both are equal as same P.d

2. *Predict* how the initial brightness of the bulb compares to the brightness of the single bulb in part A. Explain.

    Same initial brightness as both have ~~same~~ P.d equal.

3. *Predict* how the final amount of charge on the capacitor would compare to the final amount of charge on the capacitor from part A.

    Final charge on Capacitor in part E is less than in part A. Due to a smaller capacitor is used.

    Set up the circuit and check your predictions. If your prediction is in conflict with your observation, how can you account for your observation?

    with predictions even though capacitor on pw appered the same.

III. Multiple capacitors

A bulb is connected to a battery and two capacitors as shown at right. Suppose that C_1 is less than C_2.

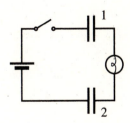

A. Before connecting the circuit a student makes the following prediction:

> *"Current flows from the positive side of the battery to the negative side of the battery. Since the bulb is isolated from the battery on both sides by the capacitors, the bulb will not light."*

Do you agree or disagree with this prediction? Explain.

Agree, clearly no current reaches it.

Obtain a second capacitor from your instructors and check your prediction.

B. Make the following predictions on the basis of your observations of this circuit. Do not use a voltmeter.

1. Just after the switch is closed:

• What is the potential difference across the bulb? Explain how you can tell from the brightness of the bulb.

P.d across bulb $V_R = V$, the bulb has high brightness.

• What is the potential difference across each of the capacitors? Explain your reasoning.

Each capacitors is zero as inital charge is zero.

$$V = \frac{q}{C} \qquad Q = 0$$
$$V = 0$$

2. A long time after the switch is closed:

• What is the potential difference across the bulb? Explain how you can tell.

Zero, as it's off so $V_R = 0$.

• What is the sum of the potential differences across the two capacitors? Explain.

$$V = V_R + V_{C1} + V_{C2} \qquad V_R = 0 \Rightarrow V = V_{C1} + V_{C2}$$

• Is the final charge on capacitor 1 *greater than*, *less than*, or *equal to* the final charge on capacitor 2? Explain.

Equal, as both capacitor charge equally.

• Is the potential difference across capacitor 1 *greater than*, *less than*, or *equal to* the potential difference across capacitor 2? Explain.

C_1 is greater than C_2 as its known $V_{C1} = \frac{q}{C_1}$ $V_{C2} = \frac{Q}{C_2}$, $C_1 < C_2 \Rightarrow V_{C1} > V_{C2}$

Use the voltmeter to check your predictions in part B.

Tyler Nappy

I. Magnetic materials

A. Investigate the objects that you have been given (magnets, metals, cork, plastic, wood, *etc.*). Separate the objects into three classes based on their interactions with each other.

1. List the objects in each of your classes.

Class 1	Class 2	Class 3
Magnets	Metals – iron – nickel – cobalt	cork Plastic Wood etc.

2. Fill out the table below with a word or two describing the interaction between members of the same and different classes.

Table of Interactions

	Class 1	Class 2	Class 3
Class 1	Attraction & Repul	only Attraction	No Interaction
Class 2	only Attraction	No Interaction	No Interaction
Class 3	No Interaction	No Interaction	No Interaction

3. Are all metals in the same class?

No, only metals as iron, nickel, cobalt are in the same class.

4. To which class do magnets belong? Are all the objects in this class magnets?

Class 1 as magnets show attraction and repulsion.

B. Obtain a permanent magnet and an object that is attracted to the magnet but not repelled. Imagine that you do not know which object is the magnet. Using only these two objects, find a way to determine which object is the permanent magnet. (*Hint:* Are there parts on either object that do not interact as strongly as other parts?)

Place the suspected magnet horizontally across from the other object to see which does and does not attract.

C. The parts of a permanent magnet that interact most strongly with other materials are called the *poles* of a magnet.

How many magnetic poles does each of your magnets have? Explain how you found them.

Two, as one side attracts and the other repels.

How many different *types* of poles do you have evidence for so far? Explain.

North and South as defining what sides attract each other.

Using three magnets, find a way to distinguish one *type* of pole from another.

Pushing the magnets together to see which sides attract and repel.

Tutorials in Introductory Physics
McDermott, Shaffer, & P.E.G., U. Wash.

©Prentice Hall, Inc.
First Edition, 2002

D. Describe how an *uncharged* pith ball suspended from a string can be used to test whether an object is charged.

To see if the pith ball gains a charge and repels or attracts towards a charged object.

Predict what will happen when an uncharged pith ball is brought near one of the poles of the magnet. Explain.

No interaction as the ball and magnet is uncharged.

Obtain a pith ball and test your prediction. Record your results.

Based on your observations above, *predict* what will happen when the pith ball is brought near the other pole of the magnet. Test your prediction.

Still no interaction as both are uncharged.

Is there a *net charge* on the north (or south) pole of a magnet? Explain.

North would be negative & south positive.

E. A paper clip is attached to a string and suspended from a straw. It is then placed so that it hangs inside an aluminum-foil lined cup as shown.

1. *Predict* what will happen to the paper clip when a charged rod is brought near the cup. Explain in terms of the electric field inside the foil-lined cup.

No interaction as there on the foil will have some included charge.

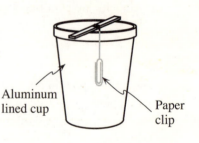

Aluminum lined cup

Paper clip

Obtain the equipment and test your prediction. Discuss this experiment with your partners.

Predict what you would observe if the paper clip were outside the cup. Explain your reasoning, then check your prediction.

It would be pushed away from the charged rod.

2. Bring a magnet near the cup and observe what happens to the paper clip inside the cup. Record your observations.

The paper clip will attract towards the magnet.

F. Based on your observations in parts D and E above, would you say that a magnetic interaction is the *same as* or *different from* an electrical interaction? Explain.

Different as other charged object doesn't affect the magnet.

II. Magnetic fields

We have observed that magnets interact even when they are not in direct contact. In electrostatics we used the idea of an electric field to account for the interaction between charges that were separated from one another. For magnetic interactions, we similarly define a *magnetic field*.

A. Obtain a compass from a tutorial instructor.

1. Use the compass to explore the region around a bar magnet.

   Describe the behavior of the compass needle both near the poles of the magnet and in the region between the poles. *It's directly along or opposite to the poles of bar magnet.*

   To which class of objects from section I does the compass needle belong? Explain. *Class 1 as it attracts and repels.*

2. Move the compass far away from all other objects. Shake the compass and describe the behavior of the compass needle. *It will denote the directions of Earth's magnetic field.*

   Does the needle behave as if it is in a magnetic field? *Yes, the North and South pole of Earth.*

   We can account for the behavior of the compass needle by supposing that it interacts with the Earth and that the Earth belongs to one of the categories from section I.

   To which class of objects from section I do your observations suggest the Earth belongs? Explain how you can tell. *Class 1 as it has a permanent magnetic field.*

3. We define the *north pole* of a magnet as the end that points toward the arctic region of the Earth when the magnet is free to rotate and is not interacting with other nearby objects.

   On the basis of this definition, is the geographic north pole of the Earth a magnetic north pole or a magnetic south pole? *A magnetic south pole with it attracting the compass.*

   Use your compass to identify the north pole of an unmarked bar magnet.

B. Place a bar magnet on an enlargement of the diagram at right.

1. Place the compass at each of the lettered points on the enlargement and draw an arrow to show the direction in which the north end of the compass points.

   Discuss with your partners how the interaction of the compass with the magnet depends on the distance from the bar magnet and the location around the bar magnet.

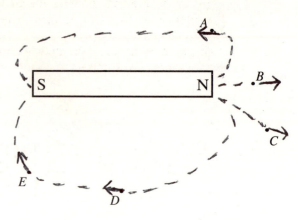

Devise a method by which you can determine the approximate relative magnitudes of the magnetic field at each of the marked locations. Explain your reasoning.

Place the compass at the marked location and identify the deflection angle.

2. We *define* the direction of the magnetic field at a point as the direction in which the north end of a compass needle points when the compass is placed at that point.

   Make the arrows on your enlargement into magnetic field *vectors* (*i.e.,* draw them so that they include information about both the magnitude and direction of the field).

   Meganetic field lines, as pointed by compass.

C. Obtain some small magnets and stack them north-to-south until you have a bar about the same length as your bar magnet. Place them on an enlargement of the diagram at right.

1. On the enlargement, sketch the magnetic field *vectors* at the locations A–E.

   How does the magnetic field of the stack of magnets compare to the magnetic field of the bar magnet?

   Both are similar like bar magnet.

2. Break the stack in half and investigate the breaking points. Describe how many north and how many south poles result. *2 north & 2 south as both are bipolar.*

   What does your observation suggest about how a bar magnet would behave when broken in half? *suggest that unipole never exist.*

Tutorials in Introductory Physics
McDermott, Shaffer, & P.E.G., U. Wash.

©Prentice Hall, Inc.
First Edition, 2002

3. On your enlargement draw the magnetic field vectors at the six locations *A–F* when just the right half of the stack of magnets is present.

Using a different color pen, draw the magnetic field vectors when just the left half of the stack of magnets is present.

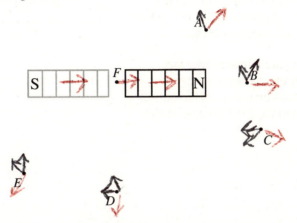

Compare the field vectors for the two half-stacks of magnets to the field vectors for the whole stack.

Is your observation consistent with the idea that magnetic fields obey the principle of superposition? Explain. It is consistent as the magnetic field outside is from North to South, while the inside is south to North.

From your observations, what can you infer about the direction of the magnetic field inside a bar magnet? Explain. It act as if the bar is made of a large number of small magnets. As field goes from N to S.

Sketch magnetic field vectors for a few points inside the magnet.

Does the magnetic field of a bar magnet always point away from the north pole and toward the south pole of the magnet? Explain.

Yes, as every time the bar breaks a new magnet with same N-S direction oppens.

What can you infer about the strength of the magnetic field inside the magnet as compared to outside the magnet? The same.

I.   The magnetic force on a current-carrying wire in a magnetic field

Obtain the following equipment:

* magnet
* wooden dowel
* ring stand and clamp
* battery
* two paper clips
* two alligator-clip leads
* 30 cm piece of connecting wire
* magnetic compass
* enlargement showing magnet and wire

Hang the connecting wire from the paper clips as shown so that it swings freely. Do not connect the wires to the battery until told to do so.

A.   On an enlargement of the figure below, sketch field lines representing the magnetic field of the bar magnet. Show the field both inside and outside the magnet.

On the diagram, indicate the direction of the current through the wire when the circuit is complete.

Predict the direction of the force exerted on the wire by the magnet when the circuit is complete. Explain.

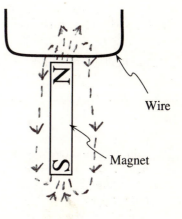

   current is from right to left as the magnetic field is stright upward as using right hand rule.

Check your prediction. (Do not leave the battery connected for more than a few seconds. The battery and wires will become hot if the circuit is complete for too long.)

B.   Make predictions for the following five situations based on what you observed in part A. Check your answers only after you have made all five predictions.

   1.   The magnet is turned so that the south pole is near the wire while the battery is connected.

   Prediction: *The magnetic field is exactly opposite.*

   Observation: *Force is now perpendicularly outside the plane.*

   2.   The leads to the battery are reversed (consider both orientations of the magnet).

   Prediction: *Current is now exactly opposite.*

   Observation: *Current would be opposite.*

Tutorials in Introductory Physics
McDermott, Shaffer, & P.E.G., U. Wash.

©Prentice Hall, Inc.
First Edition, 2002

3. The north pole of the magnet is held near the wire but the battery is not connected.

    Prediction:

    Observation: No force on wire.

4. The north pole of the magnet is held: (a) closer to the wire and (b) farther from the wire.

    Prediction: Greater force, B is stronger

    Observation: True

5. The magnet is turned so that it is parallel to the wire while the battery is connected.

    Prediction:

    Observation: No force.

Resolve any discrepancies between your predictions and your observations. (*Hint:* Consider the *vector* equation for the magnetic force on a current-carrying wire in a magnetic field: $\vec{F} = i\vec{L} \times \vec{B}$.)

II. The magnetic field of a current-carrying wire

A. Suppose you place a small magnet in a magnetic field and allow it to rotate freely. How will the magnet orient relative to the external magnetic field lines? Illustrate your answer below.

B. Suppose you hold a magnetic compass near a current-carrying wire as shown. (A *magnetic compass* is a magnet that can rotate freely.) The face of the compass is parallel to the tabletop.

    1. *Predict* the orientation of the compass needle when the circuit is complete. Sketch a diagram that shows the wire, the direction of the current through it, the direction of the magnetic field directly below the wire, and the predicted orientation of the compass needle.

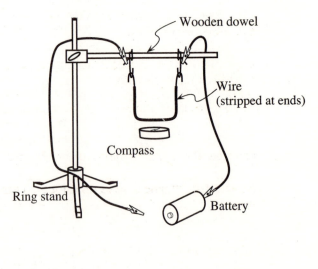

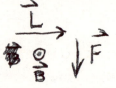

    2. Check your answer. If the deflection of the needle is not what you predicted, resolve the discrepancy. (*Hint:* Is there more than one magnetic field affecting the compass?)

     True

C. Now suppose that you hold the compass at some other locations near the wire (*e.g.,* directly above the wire or to one side of a vertical wire). For each location, *predict* the orientation of the compass needle when the circuit is closed. Make sketches to illustrate your predictions.

Check your answers. If the orientation of the compass needle is not what you predicted, resolve the discrepancy. True

D. Sketch the magnetic field lines of a current-carrying wire. Include the direction of the current in your sketch.

III. Current loops and solenoids

A. A wire is formed into a loop and the leads are twisted together. The sides of the loop are labeled A–D. The direction of the current is shown. (The diagram uses the convention that ⊙ indicates current out of the page and ⊗ indicates current into the page.)

1. On the top two diagrams at right, sketch magnetic field lines for the loop. Base your answer on your knowledge of the magnetic field of a current-carrying wire.

   Explain why it is reasonable to ignore the effect of the magnetic field from the wire leads.

   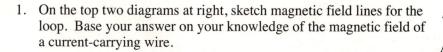

   Twisted wires outside from a solenoid.

2. Consider the magnetic field of a bar magnet.

   How are the magnetic field lines for the current loop similar to those for a short bar magnet? N & S side field lines hold same direction.

   Can you identify a "north" and a "south" pole for a current loop?

   Yes.

   Devise a rule by which you can use your right hand to identify the magnetic poles of the loop from your knowledge of the direction of the current.

Tutorials in Introductory Physics
McDermott, Shaffer, & P.E.G., U. Wash.

©Prentice Hall, Inc.
First Edition, 2002

B. A small current loop is placed near the end of a large magnet as shown.

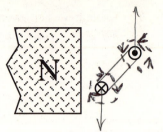

1. Draw vectors to show the magnetic force on each side of the loop.

   What is the net effect of the magnetic forces exerted on the loop?

   Attracted to magnent
   (Counter clockwise)
   countercloekwise

2. Suppose that the loop were to rotate until oriented as shown.

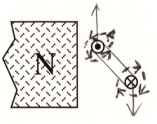

   Now, what is the net effect of the magnetic forces exerted on the loop?

   Attracts clockwise
   clockwise

   Is there an orientation for which there is no net torque on the loop? Draw a diagram to illustrate your answer.

   Yes,
   Straight up or down

3. Are your results above consistent with regarding the current loop as a small magnet? Label the poles of the current loop in the diagrams above and check your answer.

   Yes, all results keep consistent
   with what is observed.

C. A solenoid is an arrangement of many current loops placed together as shown below. The current through each loop is the same and is in the direction shown.

Obtain or draw an enlargement of the figure.

1. At each of the labeled points, draw a vector to indicate the direction and magnitude of the magnetic field. Use the principle of superposition to determine your answer.

2. Sketch magnetic field lines on the enlargement.

Describe the magnetic field near the center of the solenoid.

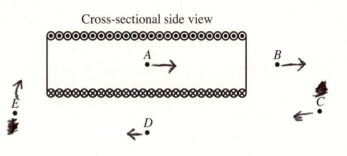

Cross-sectional side view

3. How does the field of the solenoid at points *A–E* compare with that of a bar magnet (both inside and outside)?   Essentially the same.

Which end of the solenoid corresponds to a north pole? Which end corresponds to a south pole?

North pole is right, while South pole is left.

4. How would the magnetic field at any point within the solenoid be affected by the following changes? Explain your reasoning in each case.

   *   The current through each coil of the solenoid is increased by a factor of two.

$B = \mu_0 n I \to$ B would double.

   *   The number of coils in each unit length of the solenoid is increased by a factor of two, with the current through each coil remaining the same.

$B = \mu_0 n I \to$ B would double

I. Induced currents

A. A copper wire loop is placed in a uniform magnetic field as shown. Determine whether there would be a current through the wire of the loop in each case below. Explain your answer in terms of magnetic forces exerted on the charges in the wire of the loop.

- The loop is stationary.

  No current

- The loop is moving to the right.

  No current

- The loop is moving to the left.

  No current

B. Suppose that the loop is now placed in the magnetic field of a solenoid as shown.

1. Determine whether there would be a current through the wire of the loop in each case below. If so, give the direction of the current. Explain in terms of magnetic forces exerted on the charges in the wire of the loop.

   - The loop is stationary.

     No current

   - The loop is moving toward the solenoid.

     Goes aginst negative

   - The loop is moving away from the solenoid.

     Goes opposite

2. For each case above in which there is an induced current, determine:

   - the direction of the *magnetic moment* of the loop. (*Hint:* Find the direction of the magnetic field at the center of the loop due to the induced current in the loop. The magnetic moment is a vector that points in this same direction.)

   - whether the loop is *attracted toward* or *repelled from* the solenoid.

     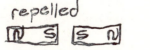

     repelled   N S S N   *attracted*   S N S N

   - whether the force exerted on the loop tends to *increase* or to *decrease* the relative motion of the loop and solenoid.

     It decrease

Tutorials in Introductory Physics
McDermott, Shaffer, & P.E.G., U. Wash.

©Prentice Hall, Inc.
First Edition, 2002

C. In each of the diagrams below, the position of a loop is shown at two times, t_o and $t_o + \Delta t$. The loop starts from rest in each case and is displaced to the right in Case A and to the left in Case B. On the diagrams indicate:

- the direction of the induced current through the wire of the loop,
- the magnetic moment of the loop,
- an area vector for each loop,
- the sign of the flux due to the external magnetic field (at both instants), and
- the sign of the induced flux (at both instants).

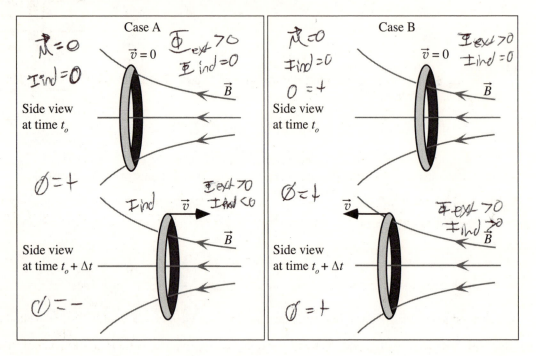

D. State whether you agree or disagree with each of the students below. If you agree, explain why. If you disagree, cite a specific case for which the student's statement does not give the correct answer. (*Hint:* Consider cases A and B above.)

Student 1: *"The magnetic field due to the loop always opposes the external magnetic field."*

I disagree, as in case B at $t_o + \Delta t$ the voltage is moved.

Student 2: *"The flux due to the loop always has the opposite sign as the flux due to the external magnetic field."*

I disagree, as in case B at $t_o + \Delta t$ flux is the same.

Student 3: *"The flux due to the loop always opposes the change in the flux due to the external magnetic field."*

I agree, as case A $\Delta I_{ext} > 0$ and $\mathcal{E}_{ind} < 0$
case B $\Delta I_{ext} < 0$ and $I_{ind} > 0$

⇨ Before continuing, check your answers to parts C and D with a tutorial instructor.

Tutorials in Introductory Physics
McDermott, Shaffer, & P.E.G., U. Wash.

©Prentice Hall, Inc.
First Edition, 2002

II. Lenz' law

A. The diagram at right shows a stationary copper wire loop in a uniform magnetic field. The magnitude of the field is *decreasing* with time.

1. Would you predict that there would be a current through the loop:

   • if you were to use the idea that there is a magnetic force exerted on a charge moving in a magnetic field? Explain your reasoning.

   No, since $q\vec{v}\cdot\vec{B}=0$ since $\vec{v}=0$

   • if you were to use the reasoning of the student in part D of section I with whom you agreed? Explain.

   Yes, as $\Delta I_{ext}<0$ in case B and $I_{ind}<0$ in case A.

2. It is *observed* that there is an induced current through the wire loop in this case. Use the appropriate reasoning above to find the direction of the current through the wire of the loop.

   The current would result in the loop becoming this ⟲→$\vec{B}$

To understand the interaction between the wire loops and solenoids in section I, we can use the idea that a force is exerted on a charged particle moving in a magnetic field. In each of those cases there was an induced current when there was relative motion between the solenoid and the wire loop. In other situations such as the one above, however, there is an induced current in the wire loop even though there is no relative motion between the wire loop and the solenoid. There is a general rule called *Lenz' law* that we can use in *all* cases to predict the direction of the induced current.

B. Discuss the statement of Lenz' law in your textbook with your partners. Make sure you understand how it is related to the statement by the student with whom you agreed in part D of section I.

   Both cases presented different answers in the solenoid and wire loop.

Tutorials in Introductory Physics
McDermott, Shaffer, & P.E.G., U. Wash.

©Prentice Hall, Inc.
First Edition, 2002

C. A wire loop moves from a region with no magnetic field into a region with a uniform magnetic field pointing into the page.

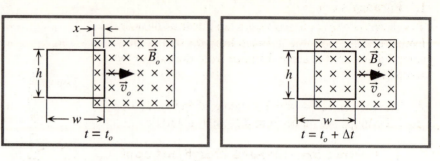

The loop is shown at two instants in time, $t = t_o$ and $t = t_o + \Delta t$.

1. Is the magnetic flux through the loop due to the external field *positive, negative,* or *zero:*

    a. at $t = t_o$? Negative

    b. at $t = t_o + \Delta t$? Negative

2. Is the *change* in flux due to the external field in the interval Δt *positive, negative,* or *zero?*

    Negative

3. Use Lenz' law to determine whether the flux due to the induced current in the loop is *positive, negative,* or *zero.*

    Positive

4. What is the direction of the current in the loop during this time interval?

    Counter-clock wise

D. At two later instants, $t = t_1$ and $t = t_1 + \Delta t$, the loop is located as shown.

1. Use Lenz' law to determine whether the flux due to the current induced in the loop is *positive, negative,* or *zero.* Explain. zero, no current.

2. Describe the current in the loop during this time interval.

    No current presented.

3. Consider the following student dialogue:

    Student 1: *"The sign of the flux is the same as it was in part C. So the current here will also be counter-clockwise."*

    Student 2: *"I agree. If I think about the force on a positive charge on the leading edge of the loop, it points towards the top of the page. That's consistent with a counter-clockwise current."*

    Do you agree with either student? Explain. No, as there is no current projected.

I. Faraday's law

Two loops of the same radius are held near a solenoid. Both loops are the same distance from the end of the solenoid and are the same distance from the axis of the solenoid.

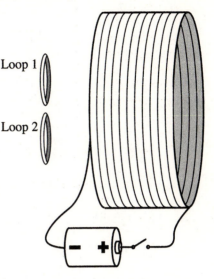

A. The resistance of loop 2 is greater than that of loop 1. (The loops are made from different materials.)

1. Is there a current induced through the wire of either of the loops:

    • before the switch is closed? Explain.

    No, due to no magnetic field or flux being presented.

    • just after the switch is closed? Explain.

    Yes, due to the flux and magnetic field being present.

    • a long time after the switch is closed? Explain.

    No, as the current, magnetic field, and flux becoming absent.

2. For the period of time that there is a current induced through the wire of the loops, find the direction of the current.

    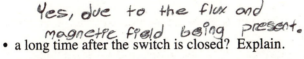 *clock wise*

3. The ratio of the induced currents for the two loops is found by experiment to be equal to the inverse of the ratio of the resistances of the loops.

    What does this observation imply about the ratio of the induced *emf* in loop 1 to the induced *emf* in loop 2? *Equal emf in both loops with current being the sole dependent on resistance.*

B. Suppose that loop 2 were replaced by a wooden loop.

    • Would there still be an emf in the loop?

    Yes

    • Would there still be a current induced in the wood loop?

    No, infinite R

C. Suppose that loop 2 were removed completely. Consider the circular path that the wire of loop 2 used to occupy.

    • Would there still be an emf along the path? Explain.

    Yes, as there would be a potential difference.

    • Would there still be a current along the path? Explain.

    No, there is no charge carrier.

Tutorials in Introductory Physics
McDermott, Shaffer, & P.E.G., U. Wash.

©Prentice Hall, Inc.
First Edition, 2002

The results of the previous exercises are consistent with the idea that a change in the magnetic flux through the surface of a loop results in an *emf* in that loop. If there is a conducting path around the loop (*e.g.*, a wire), there will be a current. The emf is independent of the material of which the loop is made; the current is not. It is found by experiment that the induced emf is proportional to the rate of change of the magnetic flux through the loop. This relationship is called *Faraday's law*. The direction of any induced current is given by Lenz' law.

D. Three loops, all made of the same type of wire, are placed near the ends of identical solenoids as shown. The solenoids are connected in series. Assume that the magnetic field near the end of each of the solenoids is uniform.

Loop 2 consists of two turns of a single wire that is twice as long as the wire used to make loop 1. Loop 3 is made of a single wire that is half as long as the wire used to make loop 1.

Just after the switch has been closed, the current through the battery begins to increase. The following questions concern the period of time during which the current is increasing.

1. Let $\mathcal{E}$ represent the induced emf of loop 1. Find the induced emf in each of the other loops in terms of $\mathcal{E}$. Explain your reasoning.

$$\mathcal{E}_2 = 2\mathcal{E}_1$$
$$\mathcal{E}_3 = \tfrac{1}{4}\mathcal{E}_1$$

Loop 1

Single loop of radius r

I

2. Let R represent the resistance of loop 1. Find the resistance of each of the other loops in terms of R. Explain.

$$R_2 = 2R_1$$
$$R_3 = \tfrac{1}{2}R_1$$

Loop 2

Double loop of radius r made from a single wire

I

3. Find the current induced through the wire of each of the loops in terms of $\mathcal{E}$ and R.

$$I_2 = I_1$$
$$I_3 = \tfrac{1}{2}I_1$$

Loop 3

Single loop of radius $r/2$

I

Tutorials in Introductory Physics
McDermott, Shaffer, & P.E.G., U. Wash.

©Prentice Hall, Inc.
First Edition, 2002

II. Applications

A. Galvanometer

Obtain a device like the one shown below. It contains a coil made of many loops of wire and a magnet suspended so that it is free to swing. A pointer has been attached to the magnet so that a small swing of the magnet will result in a large deflection of the pointer. When there is no current through the coil, the magnet is horizontal and the pointer is vertical.

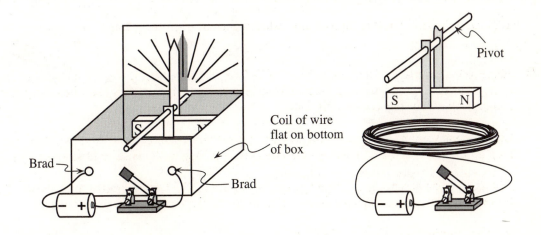

Predict the deflection of the pointer (if any) when the switch is closed. Explain the reasoning you used to make your prediction.

> The pointer will move left as to be opposite from the charge.

Connect the circuit and observe the deflection of the pointer. If your observation is in conflict with your prediction, discuss your reasoning with a tutorial instructor.

> The pointer did move to the left (cw) as the current with charge.

The device above is called a *galvanometer* and can be used to detect current. If the scale on the galvanometer has been calibrated to measure amperes, the device is called an *ammeter*.

Tutorials in Introductory Physics
McDermott, Shaffer, & P.E.G., U. Wash.

©Prentice Hall, Inc.
First Edition, 2002

B. Simple electric motor

Obtain the equipment illustrated at right and assemble it as shown.

You should have:

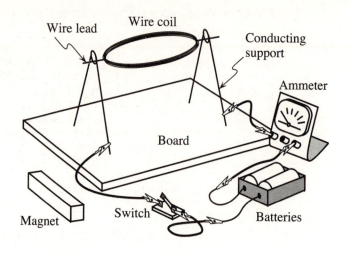

• a magnet, a battery, a switch, some connecting wire, and an ammeter.

• a copper wire coil. The ends of the wire leads to the coil have been stripped of the insulating enamel coating so that half the wire is bare.

• two conducting supports for the leads to the coil.

1. Examine the leads to the wire coil closely, so that you understand which portion of the wire has been stripped of the insulating coating.

    For what orientations of the coil will there be a current through it due to the battery?

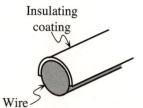

Wire lead half stripped of enamel

The wire with out the enamal would half to be touching the supports.

Check your answer by closing the switch and observing the deflection of the ammeter as you rotate the coil manually through one complete revolution.

Current only goes through without enamel.

2. Hold one pole of the magnet near the coil. Close the switch. If the coil does not begin to spin, adjust the location of the magnet or gently rotate the coil to start it spinning.

    Use the ideas that we have developed in this and previous tutorials to explain the motion of the wire coil. (The questions that follow may serve as a guide to help you develop an understanding of the operation of the motor.)

    The charge will move the the wire in the opposite direction.

Tutorials in Introductory Physics
McDermott, Shaffer, & P.E.G., U. Wash.

©Prentice Hall, Inc.
First Edition, 2002

3. When the coil is in the position shown, there is a current, *I*, through it.

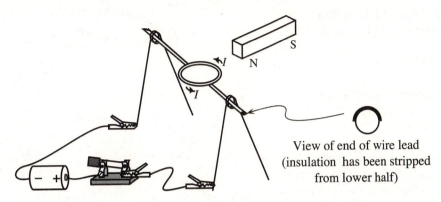

View of end of wire lead
(insulation has been stripped
from lower half)

a. The coil is manually started spinning so that it rotates clockwise.

During which portions of the cycle does the coil form a complete circuit with the battery such that there is a current through the wire of the coil?

When the wire stripped of enamel makes contact with the supports.

The current results in a magnetic moment that interacts with the magnetic field of the magnet. Will the interaction tend to *increase* or to *decrease* the angular speed of the coil? Explain.

Increase as the charge will become larger.

b. The coil is manually started spinning so that it rotates counterclockwise:

During which portions of the cycle does the coil form a complete circuit with the battery so that there is a current through the wire of the coil?

When the open wire touches the supports.

The current results in a magnetic moment that interacts with the magnetic field of the magnet. Will the interaction tend to *increase* or to *decrease* the angular speed of the coil? Explain.

Increase as the charge becomes larger.

Check that the behavior of your motor is consistent with your answers.

4. Consider the following questions about the motor:

    • Why was insulated wire used for the coil? Would bare wire also work? Explain.

    No, the insulated wire wouldn't over load the circuit.

    • Would you expect the motor to work if the leads to the coil were stripped completely? Explain. *It would work, but not as efficiently as with an insulated.*

5. Predict the effect on the motor of (i) reversing the leads to the battery and (ii) reversing the orientation of the magnet.

    (i) the motor will still run.

    (ii) the motor will spin in the opposite direction.

    Check your predictions.

Tutorials in Introductory Physics
McDermott, Shaffer, & P.E.G., U. Wash.

©Prentice Hall, Inc.
First Edition, 2002

C. Electric generator

Remove the battery and ammeter
from the circuit in part B and
insert a micro-ammeter as shown.

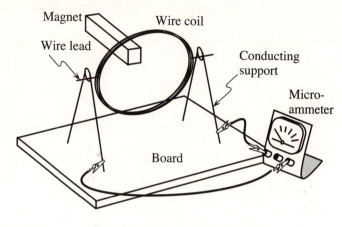

1. Suppose that the coil is made
    to spin by an external agent
    such as yourself.

    Predict the deflection of the
    micro-ammeter needle
    throughout a complete
    revolution of the coil.

The needle will move to the right.

How would your prediction change if:

• the coil were made to rotate the other way?

The needle will still move to the right.

• the poles of the magnet were reversed?

The needle will move to the left.

2. Check your predictions by gently rotating the coil so that it spins for a little time on its
    own before coming to a stop.

The second prediction was wrong as ne needle seem to Jolt left before moving right.

When the coil of the apparatus above is made to spin by an external agent, the apparatus is
called an *electric generator*.

Tutorials in Introductory Physics
McDermott, Shaffer, & P.E.G., U. Wash.

©Prentice Hall, Inc.
First Edition, 2002

Waves

I. Pulses on a spring

A tutorial instructor will demonstrate various pulses on a stretched spring. Observe the motion of the pulse and of the spring in each case and discuss your observations with your classmates.

A. A piece of yarn has been attached to the spring. How did the motion of the yarn compare to the motion of the pulse for each type of pulse that you observed?

The terms *transverse* or *longitudinal* are often used to describe the types of pulses you have observed in the demonstration. To what feature of a pulse do these terms refer?

For the rest of this tutorial we will focus on *transverse* pulses along the spring.

B. During the demonstration, did any of the following features change *significantly* as a pulse moved along the spring? (Ignore what happens when a pulse reaches the end of the spring.)

- the amplitude of the pulse

- the width of the pulse

- the shape of the pulse

- the speed of the pulse

C. During the demonstration, each of the following quantities was changed. Did any of the changes significantly affect the speed of the pulse? If so, how?

- the tension (*e.g.*, by stretching the spring to a greater length)

- the amplitude of the pulse

- the width of the pulse

- the shape of the pulse

Tutorials in Introductory Physics
McDermott, Shaffer, & P.E.G., U. Wash.

©Prentice Hall, Inc.
First Edition, 2002

II. Superposition

A. The snapshots below show two pulses approaching each other on a spring. The pictures were taken at equal time intervals. The pulses are on the "same side" of the spring (*i.e.,* each displaces the spring toward the top of the page).

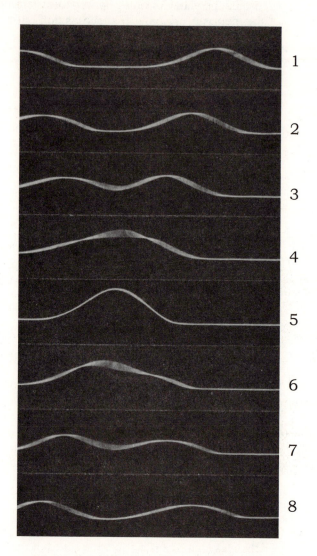

1. When the pulses meet, does each pulse continue to move in the direction it was originally moving, or does each reverse direction?

   Give evidence from the photos to support your answer.

2. When the pulses completely overlap, as shown in snapshot 5, how does the shape of the disturbance in the spring compare to the shapes of the individual pulses?

3. Describe how you could use the *principle of superposition* to determine the shape of the spring at any instant while the pulses "overlap."

4. Two pulses (1 and 2) approach one another as shown. The bottom diagram shows the location of pulse 1 a short time later.

   In the space at right, sketch the location of pulse 2 at this later time. On the same diagram, sketch the shape of the spring at this instant in time.

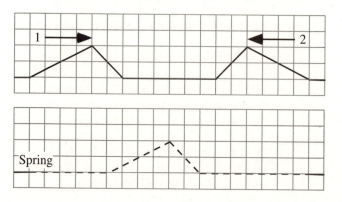

B. Two pulses of equal width and equal amplitude approach each other on *opposite sides* of a spring (*i.e.*, the pulses displace the spring in opposite directions). The snapshots below were taken at equal time intervals.

1. Is the behavior of the spring consistent with the principle of superposition? If so, which quantity is "added" in this case? If not, explain why not.

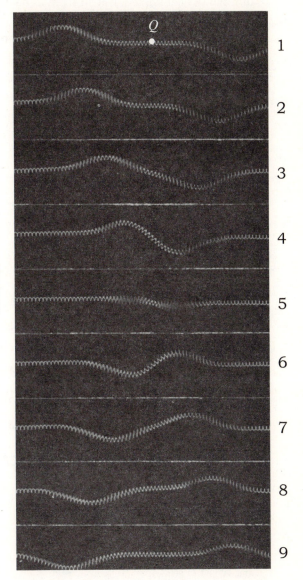

2. Below is a simplified representation of both individual pulses at a time between the instants shown in snapshots 4 and 5.

   Sketch the shape of the spring at the instant shown.

3. Let point Q be the point on the spring midway between the pulses, as shown.

   Describe the motion of point Q during the time interval shown.

4. Which, if any, of the following changes would affect the motion of point Q? Explain.

   • doubling the amplitude of both pulses

   • doubling the amplitude of just one pulse

   • doubling the width of just one pulse

5. Consider an *asymmetric* pulse as shown.

   What shape would a second pulse need to have in order that point Q not move as the two pulses pass each other?

   On the diagram, indicate the shape, location, and direction of motion of the second pulse at the instant shown.

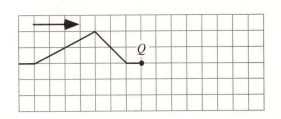

III. Reflection

A. Reflection from a fixed end

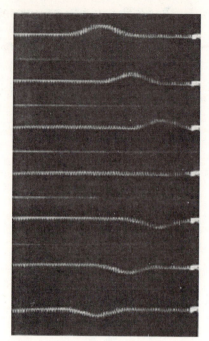

The snapshots at right show a pulse being *reflected* from the end of a spring that is held fixed in place.

1. Describe the similarities and differences between the *incident pulse* (the pulse moving toward the fixed end) and the *reflected pulse*.

2. Consider the situation in part B of section II, in which two pulses on opposite sides of a spring meet. Use a piece of paper to cover the right half of those photographs so that the portion of spring *to the left of point Q* is uncovered.

    How does the behavior of the uncovered portion of spring (including point *Q*) compare to the behavior of the spring shown at right?

The results of the exercise above suggest a model for the reflection of pulses from fixed ends of springs. We imagine that the spring extends past the fixed end and that we can send a pulse along the imaginary portion toward the fixed end. We choose the shape, orientation, and location of the imagined pulse so that as it passes the incident pulse, the end of the spring *remains fixed*. (Such a condition that governs the behavior of the end of the spring is an example of a *boundary condition*.) In this case, the reflected and imagined pulses have the same shape and orientation.

3. A pulse with speed 1.0 m/s is incident on the fixed end of a spring.

    Determine the shape of the spring at (a) *t* = 0.2 s, (b) *t* = 0.4 s, and (c) *t* = 0.6 s.

    How does the shape of the reflected pulse compare to that of the incident pulse?

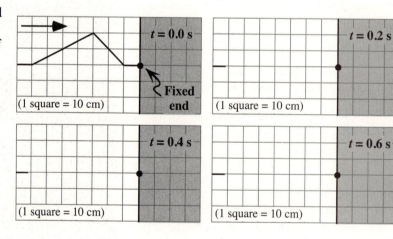

B. Reflection from a free end

Before you leave class, observe a demonstration of a pulse reflecting from the *free end* of a spring. Record your observations. You will investigate this situation in the homework.

I. Reflection and transmission at a boundary

The photographs below illustrate the behavior of two springs joined end-to-end when a pulse reaches the boundary between the springs. The snapshots were taken at equal time intervals.

A. Describe what happens after the pulse reaches the boundary between the springs.

Compare the widths of the incident and transmitted pulses.☐

B. Compare the speed of a pulse in one spring to the speed of a pulse in the other spring. Make this comparison in two ways:

1. Use the information contained in two or more snapshots. Explain.

2. Use the information contained in only a single snapshot (*e.g.*, snapshot 8). Explain.

C. In answering the questions below, assume that each spring has approximately uniform tension.

1. How does the tension in one spring compare to the tension in the other spring? Explain.

2. How does the linear mass density, μ, of one spring compare to the linear mass density of the other? Explain.

⇨ Check your answers with a tutorial instructor.

Tutorials in Introductory Physics
McDermott, Shaffer, & P.E.G., U. Wash.

©Prentice Hall, Inc.
First Edition, 2002

II. Transmission of multiple pulses

Imagine that two identical pulses are sent toward the boundary between the two springs from section I, as illustrated below. For this part of the tutorial, ignore reflected pulses.

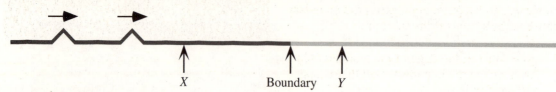

X Boundary Y

A. Imagine that you measure the time interval that starts when the crest of the first pulse reaches point X and ends when the crest of the second pulse reaches that same point. Also imagine that one of your partners measures the corresponding time interval for the transmitted pulses at point Y.

Would the time interval for the incident pulses (at point X) be *greater than, less than,* or *equal to* the time interval for the transmitted pulses (at point Y)? (*Hint:* Imagine a third person measuring this time interval at the boundary.)

Would the distance between transmitted crests be *greater than, less than,* or *equal to* the distance between incident crests? Explain.

B. Is the time it takes a single incident pulse to pass by point X *greater than, less than,* or *equal to* the time it takes a single transmitted pulse to pass by point Y?

Explain how the change in the width of the pulse as it passes from the first spring to the second is a direct consequence of the difference in speed in the two springs.

On the diagram above, sketch the transmitted pulses showing the widths and spacing of the transmitted pulses relative to the incident pulses.

⇨ Check your answers for parts A and B with a tutorial instructor.

Tutorials in Introductory Physics
McDermott, Shaffer, & P.E.G., U. Wash.

©Prentice Hall, Inc.
First Edition, 2002

III. Reflection and transmission at a boundary revisited

The springs in the photograph at right are the same as in the photographs on the first page. However, now a pulse approaches the boundary between the springs from the right.

A. After the trailing edge of the incident pulse has reached the boundary, will there be a reflected pulse?

> *If so:* On which side of the spring will the reflected pulse be located? How will its width compare to the width of the incident pulse?

> *If not:* Explain why not.

How will the transmitted pulse compare to the incident pulse?

In the space below the photograph, make a sketch that shows the shape of the springs at an instant after the incident pulse is completely transmitted. Your sketch should illustrate the relative widths of the pulse(s) and their relative distance(s) from the boundary as well as which side of the spring each pulse is on.

B. Ask a tutorial instructor for the time sequence of photographs that illustrates this situation so that you can check your predictions.

If your prediction was incorrect, identify those parts of your prediction that were wrong.

IV. A model for reflection at a boundary

We have observed that reflection occurs when a pulse reaches the boundary between two springs, that is, where there is an abrupt *change* in medium. We would like to be able to predict whether the boundary will act more like a fixed end or more like a free end.

A. In the situation illustrated in section I, are the incident and reflected pulses on the *same* side of the spring, or are they on *opposite* sides of the spring?

On the basis of this observation, does it appear that the reflection at the boundary is more like reflection from a fixed end or a free end?

Tutorials in Introductory Physics
McDermott, Shaffer, & P.E.G., U. Wash.

©Prentice Hall, Inc.
First Edition, 2002

B. Which of the following quantities are different on the two sides of the boundary?

- tension

- linear mass density

- wave speed

Which of the above quantities could you use to predict whether the boundary will act more like a fixed end or more like a free end? (It may help to consider limiting cases, *i.e.,* very large or very small values of the properties.)

Describe how you could predict whether the reflected pulse will be on the same side of the spring as the incident pulse or whether it will be on the opposite side.

Describe how you could predict whether the transmitted pulse will be on the same side of the spring as the incident pulse or whether it will be on the opposite side.

C. Imagine that a pulse on a spring is approaching a boundary. Would the boundary act more like a fixed end or more like a free end if the spring is connected to:

- a very massive chain?

- a very light fishing line?

I.   Water waves passing from shallow water to deep water

A.   The diagram at right shows a large
     tank of water containing two regions
     of different depths.

     A periodic wave is being generated at
     the left side of the tank. At the
     instant shown, the wave has not yet
     reached the deeper water. (The lines
     in the diagram, called *wavefronts,*
     represent the *crests* of the wave.)

     It is observed that water waves travel
     more quickly in deep water than in
     shallow water.

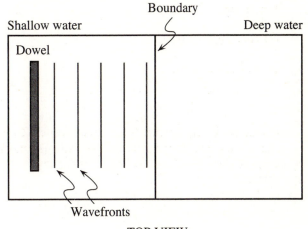

TOP VIEW

Make the following predictions based on what you have learned about the behavior of pulses
passing from one spring to another.

1.   Predict how the wavelengths of the waves in the two regions will compare. Explain.

2.   Will a crest be transmitted as *a crest, a trough,* or *something in between?* Explain.

3.   Predict how the frequencies of the waves in the two regions will compare. Explain.

⇨   Check your predictions with a tutorial instructor.

B.   Suppose that the dowel were oriented as shown and rocked back
     and forth at a steady rate. (Only part of the tank is shown.)

     On the diagram, (1) sketch the location and orientation of
     several wavefronts generated by the dowel, and (2) draw an
     arrow to show the direction of propagation of the wavefronts.

     Ask a tutorial instructor for equipment that you can use to check
     your answer experimentally. (Generate a periodic wave by
     gently rocking the dowel back and forth at a steady rate.) If
     your answer was incorrect, resolve the inconsistency.

     On the basis of your observations, how is the orientation of a
     straight wavefront related to its direction of propagation?

          Explain how your answer can apply also to circular wavefronts (such as those made by a
          drop of water falling into a tank of water). Make a sketch of circular wavefronts to
          justify your answer.

It is useful to represent straight wavefronts by drawing a single line along the direction that the wave moves. An arrowhead on the line (——▶——) indicates the direction of propagation. The line and arrowhead together are called a *ray,* and a diagram in which waves are represented by rays is called a *ray diagram.*

C. On the diagram in part B, draw a ray that shows the direction of propagation of the wave generated by the dowel.

D. Suppose that the dowel and the boundary between the shallow and deep water were oriented as shown.

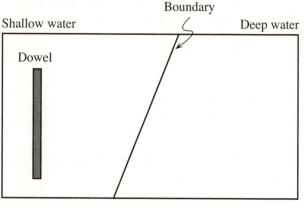

On the basis of your observations thus far, sketch two consecutive crests (1) before they cross the boundary, (2) as they are crossing the boundary, and (3) after they have crossed the boundary. (Ignore reflections at the boundary.)

Explain the reasoning you used in making your sketches.

E. Obtain a photograph that shows wavefronts incident from the left on a boundary between two regions of water and check your answers in part D.

1. Explain how you can tell from the photograph that the region of shallower depth is on the left-hand side of the photograph.

2. Describe how the wavefronts change in crossing the boundary. Use your answer to part B to determine how, if at all, the direction of propagation changes.

3. How does the phase of the wave change, if at all, in passing from one region to the other? (In other words, is a crest transmitted as *a crest, a trough,* or *something in between?*) Explain how you can tell from the photograph.

4. Are your predictions in part D consistent with your answers to the above questions? If not, resolve any inconsistencies.

Tutorials in Introductory Physics
McDermott, Shaffer, & P.E.G., U. Wash.

©Prentice Hall, Inc.
First Edition, 2002

II. A water wave passing from deep water to shallow water

A. The diagram at right shows a periodic wave incident on a boundary between deep and shallow water. Assume that the wave speed in the shallow water is *half* as great as in the deep water.

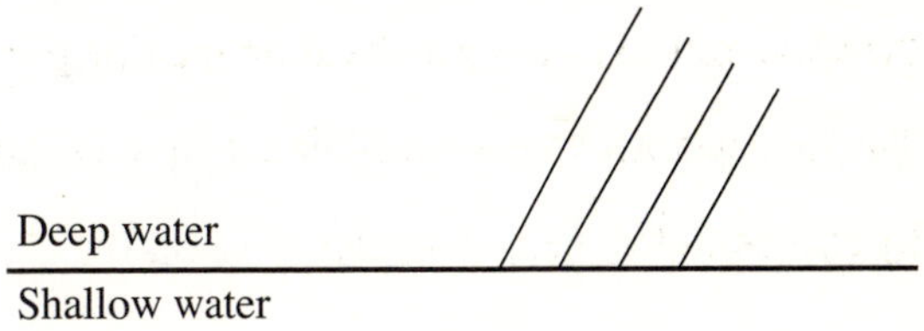

Deep water

Shallow water

Ask a tutorial instructor for an enlargement of the diagram and several transparencies.

1. Choose the transparency in which the parallel lines best represent the transmitted wavefronts.

   Explain the reasoning that you used to determine which set of parallel lines best represents the transmitted wave.

2. Place the transparency that you chose on the enlargement so that the parallel lines show the orientation and locations of the transmitted wavefronts.

   What criteria did you use to determine how to orient the transmitted wavefronts?

   Is there more than one possible orientation for the transmitted wavefronts that is consistent with your criteria?

3. Describe how the diagram would differ if the snapshot had been taken a quarter period later. (*Hint:* What is the direction of propagation of the transmitted wavefronts? How far do they travel in a quarter period?)

B. Sketch two diagrams below that illustrate waves passing from deep to shallow water at the angle of incidence shown. In one diagram, show the wavefronts; in the other, the rays.

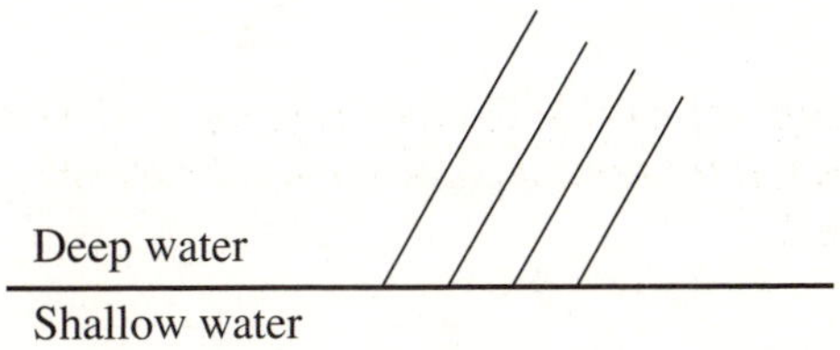

Deep water

Shallow water

Wavefront diagram

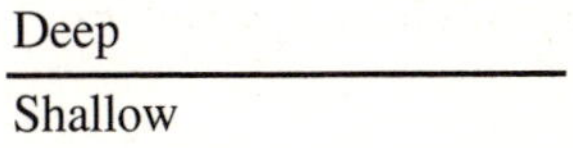

Deep

Shallow

Ray diagram

©Prentice Hall, Inc.
First Edition, 2002

The change that occurs in a wave when it propagates into a region with a different wave speed is called *refraction*. When representing waves by a ray diagram, the *angle of incidence* is defined as the angle between the ray that represents the incident wave and the *normal* to the boundary. The *angle of refraction* is defined analogously.

C. On the ray diagram in part B, label the angle of incidence, θ_i, and the angle of refraction, θ_r.

D. Obtain the equipment shown at right from an instructor. The parallel lines on the paper represent wavefronts incident on a boundary (indicated by the edge between the paper and cardboard). By changing the orientation of the paper, you can model different angles of incidence.

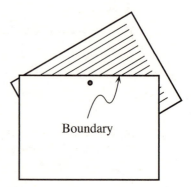

Suppose that the wavefronts on the paper are water waves in deep water approaching a region of shallow water. Rotate the paper so that the angle of incidence is 0°. Which of the transparencies used in part A can be used to represent the wavefronts in the shallow water? Place that transparency on the device to show the refracted wavefronts. Explain.

When the angle of incidence is 0°, what is the angle of refraction?

As you gradually increase the angle of incidence, does the angle of refraction *increase, decrease,* or *stay the same?*

III. Summary

A. Each of the diagrams at right shows a ray incident on a boundary between two media.

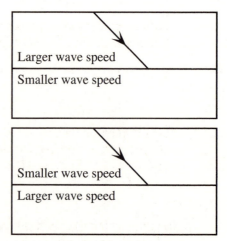

Continue each of the rays into the second medium. Using a dashed line, also draw the path that the wave would have taken had it continued without bending.

Does the ray representing a wave "bend" toward or away from the normal when:

• the wave speed is smaller in the second medium?

• the wave speed is larger in the second medium?

B. Does the ray representing a wave always "bend" when a wave passes from one medium into a different medium? If not, give an example when it does not "bend."

Tutorials in Introductory Physics
McDermott, Shaffer, & P.E.G., U. Wash.

©Prentice Hall, Inc.
First Edition, 2002

I. Representations of electromagnetic waves

A. Shown below are mathematical and pictorial representations of an electromagnetic plane wave propagating through empty space. The electric field is parallel to the z-axis; the magnetic field is parallel to the y-axis. ($\hat{x}$, $\hat{y}$, and $\hat{z}$ are unit vectors along the $+x$, $+y$, and $+z$-directions.)

$$\vec{E}(x, y, z, t) = E_o \sin(kx + \omega t)\, \hat{z} \qquad \vec{B}(x, y, z, t) = B_o \sin(kx + \omega t)\, \hat{y}$$

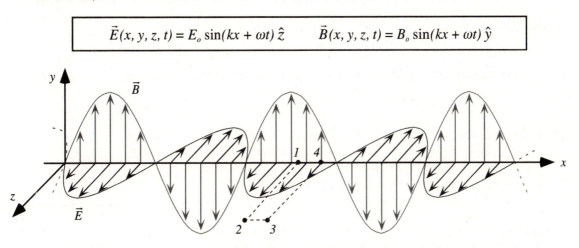

1. In which direction is the wave propagating? Explain how you can tell from the expressions for the electric field and magnetic field.

Is the wave transverse or longitudinal? Explain in terms of the quantities that are oscillating.

2. The points *1–4* in the diagram above lie in the *x–z* plane.

For the instant shown, rank these points according to the magnitude of the *electric field*. If the electric field is zero at any point, state that explicitly.

Is your ranking consistent with the mathematical expression for the electric field shown above? If not, resolve any inconsistencies. (For example, how, if at all, does changing the value of z affect the value of $\vec{E}(x, y, z, t)$?)

For the instant shown, rank points *1–4* according to the magnitude of the *magnetic field*. Check that your ranking is consistent with the expression for the magnetic field, $\vec{B}(x, y, z, t)$, above.

3. In the diagram at right, the four points labeled "×" are all located in a plane parallel to the *y–z* plane. One of the labeled points is located on the *x*-axis.

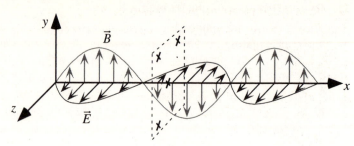

On the diagram, sketch vectors to show the direction and relative magnitude of the electric field at the labeled points.

Justify the use of the term *plane wave* for this electromagnetic wave.

⇨ Check your answers to part A with a tutorial instructor.

B. Three light waves are represented at right. The diagrams are drawn to the same scale.

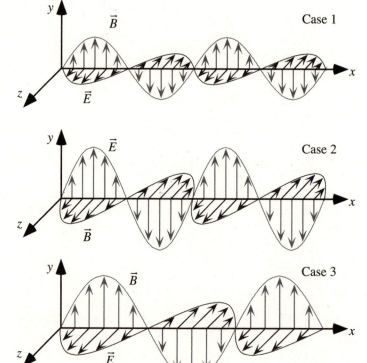

1. How is the wave in case 1 different from the wave in case 2? Explain how you can tell from the diagrams.

2. If the wave in case 2 were green light, could the wave in case 3 be *red light* or *blue light?* Explain.

II. Detecting electromagnetic waves

A. Write an expression for the force exerted on a charge, q, by (1) an electric field, $\vec{E}$, and (2) a magnetic field, $\vec{B}$.

If an electric field and a magnetic field were both present, would a force be exerted on the charge even if the charge were initially not moving? Explain.

B. Imagine that the electromagnetic wave in section I is a radio wave. A long, thin conducting wire (see figure at right) is placed in the path of the wave.

Wire

1. Suppose that the wire were oriented parallel to the z-axis.

    As the wave propagates past the wire, would the *electric field* due to the radio wave cause the charges in the wire to move? If so, would the charges move in a direction along the length of the wire? Explain.

    As the wave propagates past the wire, would the *magnetic field* due to the wave cause the charges in the wire to move in a direction along the length of the wire? Explain.

2. Imagine that the thin conducting wire is cut in half and that each half is connected to a different terminal of a light bulb. (See diagram at right.)

    Wire

    If the wire were placed in the path of the radio wave and oriented parallel to the z-axis, would the bulb ever glow? Explain. (*Hint:* Under what conditions can a bulb glow even if it is not part of a closed circuit?)

    Bulb

    How, if at all, would your answer change if the wire were oriented:

    • parallel to the y-axis? Explain.

    • parallel to the x-axis? Explain.

3. Suppose that the bulb were disconnected and that each half of the wire were connected in a circuit, as shown. (A conducting wire or rod used in this way is an example of an *antenna*.)

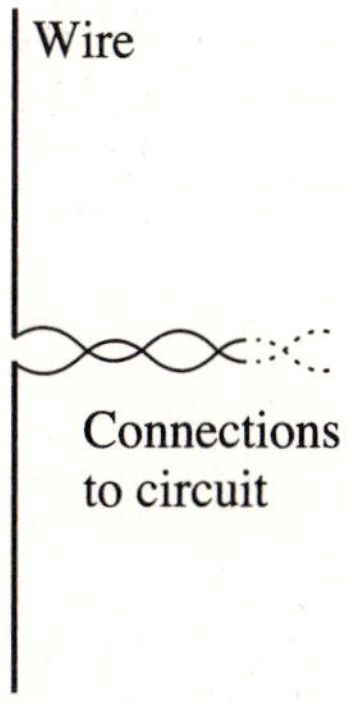

    Wire

    In order to best detect the oncoming radio wave (that is, to maximize the current through the circuit), how should the antenna be oriented relative to the wave? Explain.

    Connections
    to circuit

III. Supplement: Electromagnetic waves and Maxwell's equations

A. Recall Faraday's law, $\oint \vec{E} \cdot d\vec{l} = - \dfrac{d\Phi_B}{dt}$, from electricity and magnetism. We shall consider

how each side of the equation for Faraday's law applies to the imaginary loop,

$1 \rightarrow 2 \rightarrow 3 \rightarrow 4 \rightarrow 1$, in the figure for part A of section I.

1. For the instant shown in the figure, determine whether each quantity below is *positive, negative,* or *zero*. Explain your reasoning in each case.

   • the quantity $\int \vec{E} \cdot d\vec{l}$ evaluated over the path $1 \rightarrow 2$

   • the quantity $\int \vec{E} \cdot d\vec{l}$ evaluated over the path $2 \rightarrow 3$

   • the quantity $\oint \vec{E} \cdot d\vec{l}$ evaluated over the entire loop, $1 \rightarrow 2 \rightarrow 3 \rightarrow 4 \rightarrow 1$ (*Hint:* The answer is *not* zero!)

For an imaginary surface that is bounded by a closed loop, it is customary to use the right-hand rule to determine the direction of the area vector that is normal to that surface. For example, the vector that is normal to the flat, imaginary rectangular surface bounded by the loop $1 \rightarrow 2 \rightarrow 3 \rightarrow 4 \rightarrow 1$ points in the *positive y*-direction.

2. At the instant shown in the figure, is the magnetic flux through the loop $1 \rightarrow 2 \rightarrow 3 \rightarrow 4 \rightarrow 1$ *positive, negative,* or *zero?* Explain how you can tell from the figure.

   A short time later, will the magnetic flux through the loop be *larger, smaller,* or *the same?* Explain how you can tell from the figure.

3. According to your answers in part 2 above, is the quantity $\left(- \dfrac{d\Phi_B}{dt} \right)$, written on the right-hand side of the equation for Faraday's law, *positive, negative,* or *zero?* Explain.

   According to your results in part 1 above, is the quantity on the left-hand side of this equation *positive, negative,* or *zero?*

   Do you get the same answer for both sides of the equation for Faraday's law? If not, resolve the inconsistencies.

B. Suppose that the electric field in a light wave were $\vec{E}(x, y, z, t) = E_o \sin(kx + \omega t)\,\hat{z}$.

   Would it be possible to have a magnetic field that is *zero* for all x and t? Use Faraday's law to support your answer. (*Hint:* How, if at all, would your answers in part A above be different if the magnetic field were zero for all x and t?)

Optics

The activities in this tutorial require a darkened room. *In each experiment, make a prediction before you make any observations.* Resolve any discrepancies before continuing.

I. Light

A. Arrange a very small bulb, a cardboard mask, and a screen as shown at right. Select the largest circular hole (~1 cm in diameter) provided by the mask.

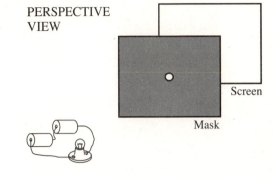

Predict what you would see on the screen. Explain in words and with a sketch.

Predict how moving the bulb upward would affect what you see on the screen. Explain.

Perform the experiments and check your predictions. If any of your predictions were incorrect, resolve the inconsistency.

B. *Predict* how each of the following changes would affect what you see on the screen. Explain your reasoning and include sketches that support your predictions.

- The mask is replaced by a mask with a *triangular* hole.

- The bulb is moved farther from the mask.

Perform the experiments and check your predictions. Resolve any inconsistencies.

C. A mask with a circular hole is placed between a bulb and a screen.

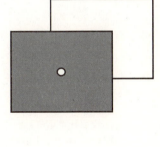

Predict how placing a second bulb above the first would affect what you see on the screen. Explain.

Predict how moving the top bulb upward slightly would affect what you see on the screen. Explain.

Perform the experiments. Resolve any inconsistencies.

D. What do your observations suggest about the path taken by light from the bulb to the screen?

Tutorials in Introductory Physics
McDermott, Shaffer, & P.E.G., U. Wash.

©Prentice Hall, Inc.
First Edition, 2002

E. Imagine that you held a string of closely spaced small bulbs one above the other. What would you expect to see on the screen?

Predict what you would see on the screen if you used a bulb with a long filament instead. Explain.

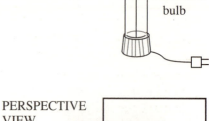

Long-filament bulb

Check your prediction.

F. The mask used in parts C–E is replaced by one that has a triangular hole as shown.

Predict what you would see on the screen when a small bulb is held next to the top of a long-filament bulb as shown. Sketch your prediction below.

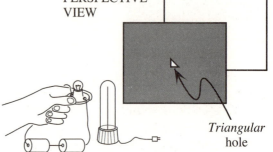

PERSPECTIVE VIEW

Triangular hole

Compare your prediction with those of your partners. After you and your partners have come to an agreement, check your prediction. Resolve any inconsistencies.

G. *Predict* what you would see on the screen in the situation pictured at right.

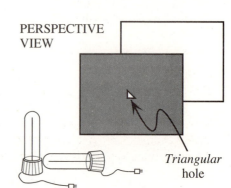

PERSPECTIVE VIEW

Triangular hole

Predict what you would see on the screen if the mask were *removed*.

Check your predictions. If any of your predictions were incorrect, resolve the inconsistency.

H. *Predict* what you would see on the screen when an ordinary frosted bulb is held in front of a mask with a triangular hole as pictured at right.

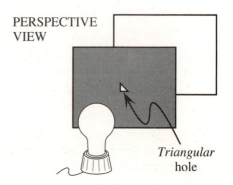

PERSPECTIVE VIEW

Triangular hole

⇨ Discuss your prediction with a tutorial instructor. Then obtain a frosted bulb and check your prediction.

Tutorials in Introductory Physics
McDermott, Shaffer, & P.E.G., U. Wash.

©Prentice Hall, Inc.
First Edition, 2002

II. Light: quantitative predictions

A. *Predict* the size of the lit region on the screen at right. Treat the bulb as a point source of light, *i.e.,* as if all the light emanates from a single point.

How would the vertical length of the lit region *change* if the diameter of the hole were halved? (In particular, would it become half as tall?) Explain.

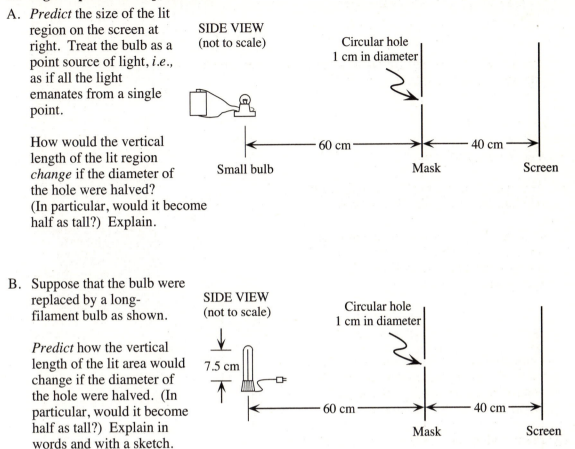

B. Suppose that the bulb were replaced by a long-filament bulb as shown.

Predict how the vertical length of the lit area would change if the diameter of the hole were halved. (In particular, would it become half as tall?) Explain in words and with a sketch.

Check your prediction. If your prediction was incorrect, resolve the inconsistency.

Predict the *approximate* height and shape of the lit region on the screen in the limit as the hole becomes very small, *e.g.,* the size of a pinhole. (*Hint:* In this limit, would the lit region be taller than, shorter than, or the same height as the filament?)

C. *Predict* what you would see on the screen in the situation pictured at right.

PERSPECTIVE
VIEW

How would the height and width of each lit region *change* if the diameter of the hole in the mask were halved? Explain.

Check your predictions. If any of your predictions were incorrect, resolve the inconsistency.

III. Supplement: Shadows

Obtain a box, thread, and small bead (~5 mm in diameter). Hang the bead as shown.

PERSPECTIVE VIEW

A. *Predict* what you would see on the screen at the back of the box in the situation pictured at right. Explain your reasoning.

Bead

Predict how placing a second bulb above the first would affect what you see on the screen. Explain your reasoning.

PERSPECTIVE VIEW

Perform the experiments. Resolve any discrepancies between your observations and predictions.

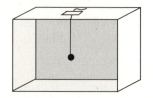

Must a region be *completely* without light for a shadow to be formed? Explain.

B. Suppose that the light from the top bulb in the situation above were red and the light from the lower bulb were green. Predict what you would see on the screen. Explain.

Obtain red and green filters from a tutorial instructor and perform the experiment described above. If your predictions were incorrect, find the error in your reasoning.

C. *Predict* what you would see on the screen in the situation shown at right. Explain your reasoning.

Suppose that the light from the vertical bulb were red and the light from the horizontal bulb were green. *Predict* what you would see on the screen.

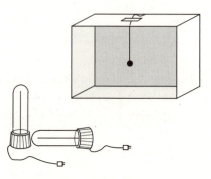

Perform the experiments described above. Resolve any discrepancies between your observations and your predictions.

I.   The method of parallax

A. Close one eye and lean down so that your open eye is at table level. Have your partner drop a very small piece of paper (about 2 mm square) onto the table.

Hold one finger above the table and then move your finger until you think it is directly above the piece of paper. Move your finger straight down to the table and check whether your finger is, in fact, directly above the paper.

Try this exercise several times, with your partner dropping the piece of paper at different locations. Keep your open eye at table level. After several tries, exchange roles with your partner.

How can you account for the fact that when your finger misses the piece of paper, your finger is always either in front of the paper or behind it, but never to the left or right of the paper?

It is easier to align the finger on a visible x-axis rather than a y-axis.

B. Suppose that you placed your finger behind the paper (as shown at right) while trying to locate the piece of paper.

Top view diagram

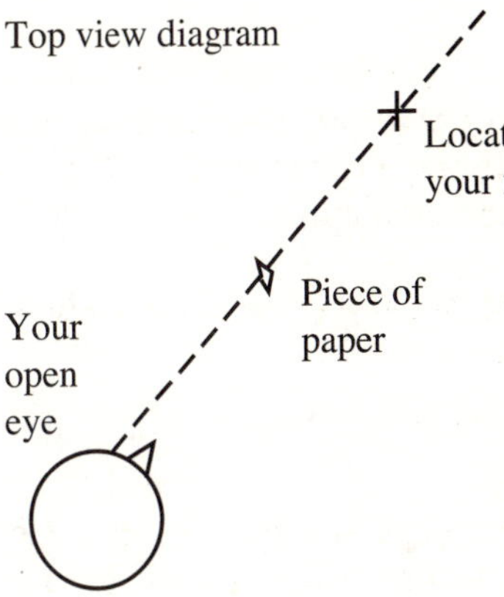

Predict whether your finger would appear to be located *to the left of, to the right of,* or *in line with* the piece of paper if:

• you moved your head to the left.

In line with paper.

• you moved your head to the right.

In line with paper.

Check your predictions. Resolve any inconsistencies.

Location of your finger

Piece of paper

Your open eye

C. Suppose that you had placed your finger in front of the piece of paper rather than behind it.

Predict whether the paper or your finger would appear on the left when you move your head to the left. Check your answer experimentally.

The finger would appear on the left as it's infront of the paper.

D. Devise a method based on your results from parts B and C by which you could locate the piece of paper. Your method should include how to tell whether your finger is directly over the piece of paper and, if not, whether it is in front of or behind the piece of paper. Describe your method to your partner, then test your method.

Align finger along x-axis, keep finger in front of paper along y-axis.

➪   Check your method with a tutorial instructor.

Tutorials in Introductory Physics
McDermott, Shaffer, & P.E.G., U. Wash.

©Prentice Hall, Inc.
First Edition, 2002

We will refer to the method that you devised for locating the piece of paper as the *method of parallax*.

II. Image location

Obtain a small mirror and two identical nails. Place the mirror in the middle of a sheet of paper. Stand one nail on its head about 10 cm from the front of the mirror. We will call this nail the *object nail*.

Top view

⬛▨▨▨▨▨▨▨▨▨⬛ Mirror

On the paper, mark the locations of the mirror and object nail.

A. Place your head so that you can see the image of the nail in the mirror.

●

Object nail

Use the method of parallax to position the second nail so that it is located in the same place as the image of the object nail. Mark this location on the paper.

Is the image of the nail located *on the surface of, in front of*, or *behind* the mirror? Explain.

Behind the mirror, as a plane mirror is virtual and behind the mirror.

Would observers at other locations agree that the image is located at the place you marked? Check your answer experimentally.

Others agree.

B. Move the nail off to the right side of the mirror as shown. Find the new image location.

Top view

⬛▨▨▨▨▨▨▨▨▨⬛ Mirror

Still behind the mirror as the concept is the same.

●

Object nail

In the following experiments, we will determine the location of an object and its image by a different technique called *ray tracing*. This technique is based on a model for the behavior of light in which we envision light being either emitted in all directions by a luminous object (such as a light bulb) or reflected in all directions by a non-luminous object (such as a nail).

III. Ray tracing

A. Place a large sheet of paper on the table. Stand a nail vertically at one end of the piece of paper.

Place your eye at table level at the other end of the piece of paper and look at the nail. Use a straightedge to draw a *line of sight* to the nail, that is, a line from your eye to the nail.

Repeat this procedure to mark lines of sight from three other very different vantage points, *then remove the nail*.

How can you use these lines of sight to determine where the nail was located?

Extend all the lines of sight further to a meeting point.

What is the smallest number of lines of sight needed to determine the location of the nail?

Two lines of sight is all thats needed.

B. Turn the large sheet of paper over (or obtain a fresh sheet of paper). Place the mirror in the middle of the sheet of paper, and place a nail in front of the mirror. On the paper, mark the locations of the mirror and the nail.

On the paper, draw several lines of sight to the *image* of the nail.

How can you use these lines of sight to determine the location of the image of the nail?

The position will be given by extending lines of sights upto where they meet.

Use the method of parallax to determine the location of the image of the nail.

Do these two methods yield the same location of the image (to within reasonable uncertainty)?

Yes both the methods will yeild the same result.

C. Remove the mirror and the object nail. For each eye location that you used in part B, draw the path that light takes from the object nail to the mirror.

Draw an arrow head on each line segment (————>) to indicate the direction that light moves along that part of the path.

On the basis of the paths that you have drawn, formulate a rule that you can use to predict the path that light takes after it is reflected by a mirror.

Virtual images will have negative distances object distance is positive.

D. Place the second nail at the location of the image of the object nail. Draw a diagram illustrating the path of the light from that nail to your eye for the same eye locations as in part C.

How is the diagram for this situation similar to the diagram that you drew in part C?

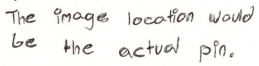

Is there any way that your eye can distinguish between these two situations?

The negative and positive location along the mirror.

IV. An application of ray tracing

In this part of the tutorial, use a straightedge and a protractor to draw rays as accurately as possible.

A. On the diagram at right, draw one ray from the pin that is reflected by the mirror.

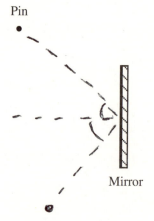

If you were to place your eye so that you were looking back along the reflected ray, what would you see?

The image location would be the actual pin.

From one ray *alone* do you have enough information to determine the location of the image? If not, what can you infer about the location of the image from only a single ray?

Not enough information, we can only infer a relative location.

B. On the diagram above, draw a second ray from the pin that is reflected by the mirror and that would reach an observer at a different location.

What can you infer about the location of the image from this second ray *alone*?

In line with the mirror.

How can you use the two rays that you have drawn to determine the location of the image?

Addust for the relative location given.

Tutorials in Introductory Physics
McDermott, Shaffer, & P.E.G., U. Wash.

©Prentice Hall, Inc.
First Edition, 2002

Is there additional information about the image location that can be deduced from three or more rays?

Additional rays give a more ~~exactly~~ precise location.

C. Determine the image location using the method of ray tracing from section III. (If it is necessary to extend a ray to show from where light appears to come, use a *dashed line*.)

Does the light that reaches the observer actually come from the image location or does this light only *appear* to come from that point?

comes from the image location.

What is the smallest number of rays that you must draw in using ray tracing to determine the location of the image of an object?

Two rays are the minimum.

How does the distance between the mirror and the image location compare to the distance between the mirror and the pin?

The distance is an estimate of the location is exact.

The diagram that you drew above to determine the image location is called a *ray diagram*. The point from which the reflected light appears to come (*i.e.*, the location of the pin that you saw when you looked in the mirror) is called the *image location*. An image is said to be *virtual* when the light that forms the image does not actually pass through the image location. An image is said to be *real* when the light that forms the image does pass through the image location.

When drawing ray diagrams, use a solid line *with an arrow head* (⟶) to represent a ray, that is, a path that light takes. Use a dashed line (- - - - - - -) to extend a ray to show from where light *appears* to come in order to distinguish such a line from an actual ray.

Tutorials in Introductory Physics
McDermott, Shaffer, & P.E.G., U. Wash.

©Prentice Hall, Inc.
First Edition, 2002

I. Cylindrical mirrors

A. A pin is placed in front of a cylindrical mirror as shown in the top view diagram at right. Lines *A–E* represent some of the light rays from the pin that reach the mirror. Points *M* and *N* represent the locations of two observers.

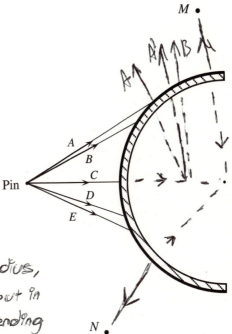

You have been provided with an enlargement of this top view diagram.

1. Use a ruler and a protractor to draw the reflected rays on the enlargement. (*Hint:* The center of the cylindrical mirror is marked on the diagram.)

    Describe how you determined the direction of each reflected ray.

    Focal length is half the radius, the rays are diverged out in different directions depending on incident rays.

2. For each of the reflected rays, use a dashed line to show the direction from which the reflected ray appears to have come.

    Do all of the reflected rays appear to have come from the same point?

    Yes, they combine at the focus point of the mirror.

3. On the diagram, draw a ray, *A'*, between rays *A* and *B*. Draw the corresponding reflected ray.

    Which more nearly appear to pass through the same point: the reflected rays *A*, *A'*, and *B* or the reflected rays *A*, *B*, and *C?*

    Reflected rays A, A', and B.

Tutorials in Introductory Physics
McDermott, Shaffer, & P.E.G., U. Wash.

©Prentice Hall, Inc.
First Edition, 2002

Describe how, in principle, you could determine the location at which an observer at *M* would see an image of the pin. Label the approximate location on the diagram.

Following the rays and eventually camying to a meeting point.

Determine and label the approximate location at which an observer at *N* would see an image of the pin.

Would the observers at *M* and *N* agree on the location of the image of the pin? Explain how you can tell from your ray diagram.

Following the rays to the mirror then from the mirror to point of interest.

4.   Ask a tutorial instructor for a semi-cylindrical mirror. Place the mirror on the enlargement and use the method of parallax to check your predictions. (You may find it helpful to tape the mirror onto the diagram.) If there are any inconsistencies between your predictions and your observations, resolve the inconsistencies.

N/A

B.   Could you use any two rays (even those that do *not* pass near a particular observer) to find the location at which that observer sees the image of the pin in the case of:

• a plane mirror? Explain.

Focal length is length is infinite, it has parallel rays, thus no distinction in image formation.

• a curved mirror? Explain.

2 rays can become a focal point where they meet, so it can be deduced easily where the observer the image.

Tutorials in Introductory Physics
McDermott, Shaffer, & P.E.G., U. Wash.

©Prentice Hall, Inc.
First Edition, 2002

C. Observers at *M* and *N* are looking at an image of the
pin in the mirror.

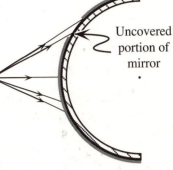

1. Suppose that all but a small portion of the mirror
were covered as shown at right.

    How, if at all, would this change affect what the
observers at *M* and *N* see? Explain.

    M and N wouldn't see
any image as being blocked.

    Determine the region in which an observer must
be located in order to see an image of the pin.
Discuss your reasoning with your partners.

    Only at rays A and B will see.

    Would two observers at different locations in this
region agree on the approximate location of the image? Explain.

    No, two observers at different location
want agree on the same location.

2. Suppose that all but a small portion of the mirror
near the center were covered, as shown at right.

    Determine the region in which an observer must be
located in order to see an image of the pin.

    Would two observers at different locations in this
region agree on the approximate location of the
image? If so, find the approximate image location.
If not, explain how you can tell.

    Yes, two observers can
agree on same approximate
image.

    Check your answers experimentally.

While the image location is independent of observer location in certain cases (*e.g.*, plane

mirrors), in general it is not. In many cases, however, it is possible to identify a limited range of

locations for which the image location is essentially independent of the observer location. An

example is when both the object and the observer lie very nearly along the axis of a cylindrical or

spherical mirror. In this situation, all rays are said to be *paraxial,* that is, they make small angles

with the axis of the mirror. Ray diagrams often specify the location of an image but not the

observer's location. For such a diagram, it should be assumed that the image location is

independent of the observer's location.

II. Multiple plane mirrors

A. Stick a pin into a piece of cardboard and place two mirrors at right angles near the pin as shown in the top view diagram below.

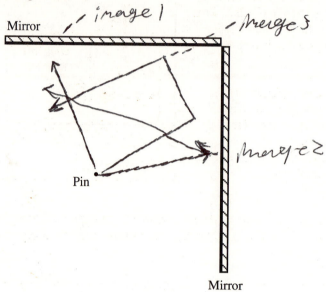

1. Describe what you observe.

   Multiple overlapping rays.

2. View the arrangement from several locations and use the method of parallax to place a pin at each of the image locations.

3. Suppose that one of the mirrors were removed. Predict which image(s) you would still see and which image(s) would vanish.

   Image 2 and 3 will disappear.

   Check your predictions. If any of your predictions were incorrect, resolve the conflict before continuing. *N/A*

4. On the diagram above, sketch a ray diagram that accounts for each image.

   Describe how one of the images differs from the others.

   They intersect one another.

B. Gradually decrease the angle between the mirrors while keeping the pin between the mirrors.

   How can you account for the presence of the additional images that you observe?

I. Image location

A. A pin is held vertically at the back of a clear square container of water as shown at right. The portion of the pin below the surface of the water is not shown.

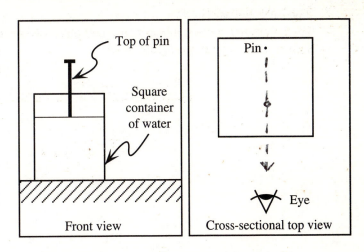

Top of pin

Square container of water

Front view

Pin

Eye

Cross-sectional top view

1. On an enlargement of the top view diagram, sketch several rays from the pin that pass through the water and out the front of the container, near the observer's eye.

For simplicity in answering the following questions, ignore the walls of the container (*i.e.*, use the approximation that light passes directly from water to air, where it travels more quickly).

- On the basis of the rays that you have drawn, *predict* where the bottom of the pin would *appear* to be located to the observer. Discuss your reasoning with your partners.

   Would appear to be closer than their real position.

- Would the bottom of the pin *appear* to be located *closer to, farther from,* or *the same distance from* the observer as the top of the pin? Explain.

   Due to the apparent depth concept the object would appear closer.

2. Obtain the necessary equipment and use the method of parallax to check your predictions. If your ray diagram is not consistent with your observations, modify your ray diagram.

   N/A   It is consistent.

The place where the pin *appears* to be located is called the location of the image of the pin, or the *image location*.

3. In part 1, you assumed that light from the pin passes directly from water to air. Devise an experiment that would allow you to test whether this approximation is valid. (*Hint:* Use the method of parallax to see how the container *alone* affects the apparent location of the pin.)

   Seems to get consistent results as in part one.

Perform this experiment and check your answer.

Tutorials in Introductory Physics
McDermott, Shaffer, & P.E.G., U. Wash.

©Prentice Hall, Inc.
First Edition, 2002

B. Suppose that a pencil were held vertically at the back of a circular beaker of water, as shown. (The portion of the pencil below the surface of the water is not shown.) *Note:* The center of the beaker is marked by an "✕."

Is the image of the bottom of the pencil *closer to, farther from,* or *the same distance from* the observer as the top of the pencil?

Sketch a qualitatively correct ray diagram to support your answer.

It will appear at the same distance.

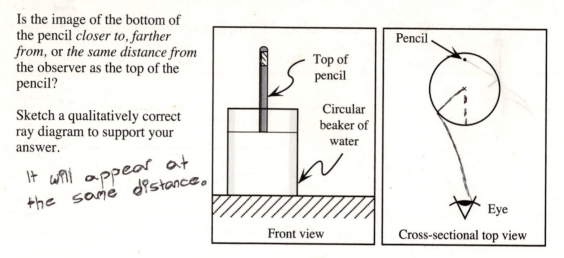

Top of pencil

Circular beaker of water

Pencil

Eye

Front view   Cross-sectional top view

Use the method of parallax to place a second pencil at the location of the image of the bottom of the pencil, and check your predictions. If your prediction was incorrect, find your error.

Prediction is correct.

C. Three students are discussing their results from part B:

Student 1:   *"I think that the image is closer to me than the pencil itself. The closer something is, the bigger it looks. Because the image of the pencil appears wider than the pencil itself, the image must be closer to me than the pencil."*

Student 2:   *"That sounds reasonable, but when I used parallax to determine the location of the image of the pencil, I found that it was farther from me than the pencil."*

Student 3:   *"That doesn't make sense though. If the image were behind the pencil, then how could I see the image? Wouldn't the pencil block my view of the image?"*

Do you agree or disagree with each of these students? Discuss your reasoning with your partners.

I disagree with each student as non states the image to be the same distance with distortion.

II. Real and virtual images

Each of the ray diagrams below illustrates the path of light from a pin through a beaker of water. In one case, the pin is near the beaker of water; in the other case, far from the beaker.

A. For each case shown below, determine and label the location of the image of the pin. Explain how you determined your answer.

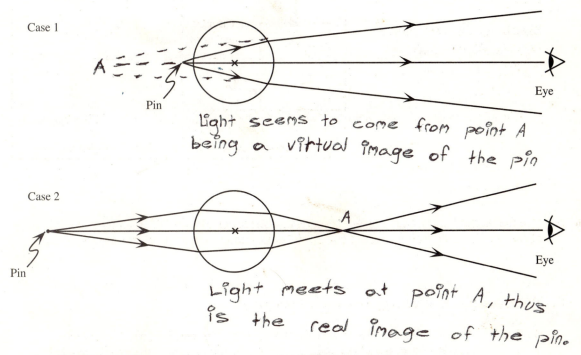

Case 1

Light seems to come from point A being a virtual image of the pin

Case 2

Light meets at point A, thus is the real image of the pin.

1. Use the ray diagrams above to answer the following questions:

   In each case, which is farther from the observer:

   • the image (below the water's surface) or → *1st Case*
   • the object (above the water's surface)? → *2nd case*

   In each case, which is farther from the observer:

   • the image (below the water's surface) or → *1st case*
   • the beaker of water? → *2nd case*

2. Use the method of parallax to check your answers for both cases 1 and 2. Resolve any inconsistencies.

   No inconsistencies.

B. In each of the previous cases, predict what would be seen on a white paper screen placed at the image location. Imagine that the room has been darkened but that the pin is illuminated. (*Hint:* In either case, does the light from the pin that forms the image pass through the image location?)

Case 1 is a virtual image and won't appear on screen.

Case 2 is a real image and would appear on screen.

Replace the pin with a lighted long-filament bulb and check your predictions. If either of your predictions were incorrect, resolve the inconsistency.

No inconsistency.

C. One of the images in part A is real; the other, virtual. Explain how you can tell which image is real and which is virtual on the basis of (1) the ray diagrams and (2) your observations in part B.

Case 1 - virtual, behind object
Case 2 - real, in front of object

Case 1 - rays never meet
Case 2 - rays meet

D. Explain how you can use a screen to determine the location of an image. In what cases, if any, would this method fail?

Case 1 will fail being a virtual image.

Case 2 - Displace the screen in such away so that a clear image is formed. The diameter of image will be minimum at a particular point. This is image position.

In this tutorial, use a straightedge to draw lines that are meant to be straight.

I. Convex lenses

A. Look at a very distant object through a convex lens. Hold the lens at arm's length so that you see a sharp image of the distant object.

Is the image formed by the lens *in front of, behind,* or *on the surface of* the lens? Use the method of parallax to determine the approximate distance between the image and the lens.

In front of the lens

B. Consider a point on the distant object that is also on the principal axis of the lens.

On the diagram below, sketch several rays from this distant point that reach the lens.

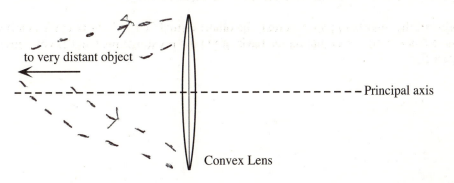

How are these rays oriented with respect to one another and to the principal axis? Explain.

They are parallel to one another.

On the basis of your observations from part A, show the continuation of each of these rays through the lens and out the other side. On the diagram, indicate where the rays converge.

Note: Refraction takes place at the two surfaces of the lens. However, in drawing a ray diagram for a thin lens, it is customary to draw rays as if all refraction takes place at the center of the lens.

C. Suppose that you placed a very small bulb at the location of the image in part B.

How would the rays from the bulb that have passed through the lens be oriented? Draw a diagram to illustrate your answer. Explain.

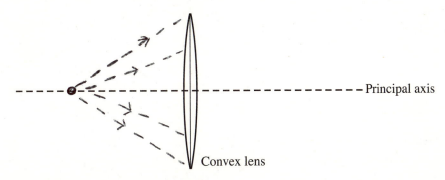

⇨ Discuss your answers with a tutorial instructor.

Tutorials in Introductory Physics
McDermott, Shaffer, & P.E.G., U. Wash.

©Prentice Hall, Inc.
First Edition, 2002

The point of intersection of the principal axis and the image of a very distant object is called the *focal point*. The distance between the center of the lens and the focal point is called the *focal length*.

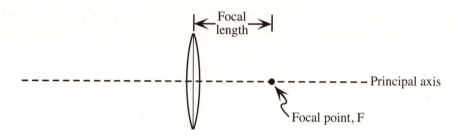

II. Ray tracing and convex lenses

The diagram below shows several rays from the eraser on a pencil that reach a convex lens.

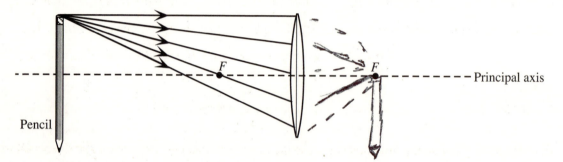

A. Consider the ray that is parallel to the principal axis.

Explain how you can use your observations from section I to draw the continuation of that ray on the right side of the lens. Draw this ray on the diagram.

The rays meet to make a image.

B. Consider the ray that goes through the focal point on the left side of the lens.

Explain how you can use your answers to part C of section I to draw the continuation of that ray on the right side of the lens. Draw this ray on the diagram.

Follow the ray back to point of refraction.

C. How can you use these two rays to determine the location of the image of the eraser? On the diagram, label the image location.

The refraction will cause an image the minimum diameter is the position.

D. Consider the ray from the eraser that strikes the lens near its center, where the sides of the lens are nearly parallel.

Using the image location as a guide, draw the continuation of this ray through the lens and out the other side.

In your own words, describe the path of a ray that passes through the center of the lens.

Refracted into two rays.

E. Draw the continuation of the two remaining rays shown on the diagram through the lens and out the other side.

The rays that you drew in parts A, B, and D are called *principal rays,* and they are useful in determining the location of an image. In some cases, one or more of these rays may not actually pass through the lens; however, they may still be used in determining the image location. The principal rays are only a few of the infinitely many that we might draw from one point on the object.

F. On the diagram on the previous page, use the three principal rays from the *tip* of the pencil to determine the location of the image of the tip of the pencil. If possible, use a different color ink or pencil for this second set of rays.

G. The diagram below shows a small object placed near a convex lens. Draw all three principal rays and determine the location of the image. Clearly label the image location.

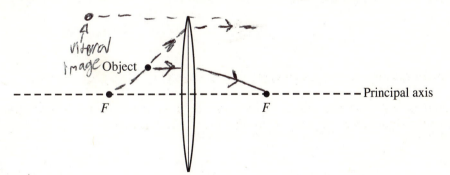

In your own words, describe how you knew to draw each ray.

Trace back all rays of light to vitural image.

III. Applications

A. A lens, a bulb, and a screen are arranged as shown below. A sharp, inverted image of the filament (not shown) appears on the screen when it is at the location shown.

Bulb   Lens   Screen

Predict how each of the following changes would affect what you see on the screen. Support your predictions with one or more ray diagrams.

• The screen is moved *closer to* or *farther from* the lens.

If closer magnification increased if farther the opposite.

• The top half of the lens is covered by a mask.

Inverted image.

Does your answer depend on which side of the lens the mask is placed? If so, how? If not, why not?

It is dependent as it can invert the image.

B. Obtain the necessary equipment and check your predictions. In the space below, record how, if at all, your predictions were different from your observations. If your predictions were incorrect, resolve the inconsistencies.

No inconsistencies.

C. If the screen were removed, would you still be able to see an image of the bulb? Does it matter where your eye is located?

If we place our eyes at the position of screen, we can observe the image.

Turn off (but do not move) the bulb, remove the screen, and check your predictions. If your predictions were incorrect, resolve the inconsistencies. (*Hint:* Was a screen necessary to see an image in earlier situations, such as the situation in part A of section I?)

Tutorials in Introductory Physics
McDermott, Shaffer, & P.E.G., U. Wash.

©Prentice Hall, Inc.
First Edition, 2002

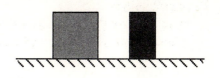

I. Apparent size

A. The diagram at right illustrates what an observer sees when looking at two boxes on a large table.

From the diagram *alone:*

• is it possible to determine which box is closer to the observer?

• is it possible to determine which box *appears* wider to the observer?

• is it possible to determine which box *actually is* wider?

Discuss your reasoning with your partners.

B. Obtain two soda cans and a cardboard tube that has a smaller diameter than the can.

1. How can you arrange the two soda cans so that (a) they *appear* to be equally wide and (b) one can *appears* wider than the other?

   In the space below, draw a top view diagram for each case that can be used to compare the apparent widths of the cans.

2. How can you arrange one can and the tube so that (a) the two objects *appear* to be equally wide and (b) the tube *appears* wider than the can?

   In the space below, draw a top view diagram for each case that can be used to compare the apparent widths of the two objects.

3. What quantities affect the apparent size of an object? Describe how increasing or decreasing each quantity affects the apparent size of that object.

   Explain how you can use a top view diagram to determine whether one object *appears* wider or narrower than another object to an observer at a particular location.

Tutorials in Introductory Physics
McDermott, Shaffer, & P.E.G., U. Wash.

©Prentice Hall, Inc.
First Edition, 2002

II. The image of an extended object

The ray diagram below shows a side view of a thin converging lens, a pencil, the image of the pencil, and five observer locations *(1–5)*. Two rays from the pencil tip are drawn through the lens.

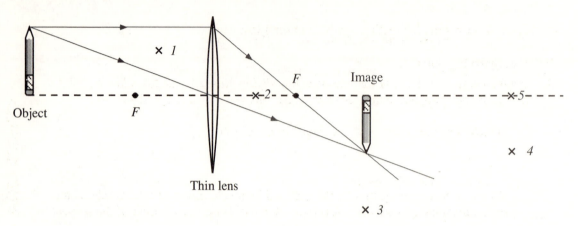

A. Could an observer at each of the labeled points see a sharp image of the pencil tip (other than the actual pencil tip)? In each case, explain why or why not. Additionally, if an observer is able to see the image, indicate the direction that the observer would have to look to see the image.

• point *1*

• point *2*

• point *3*

• point *4*

• point *5*

B. Use the above diagram to answer the following questions.

1. From which of the labeled points could an observer see the image of the *eraser?* Draw rays to support your answer. (If possible, use a different color ink to draw these rays.)

From which point(s) could an observer see the entire image of the pencil? Explain.

To an observer at such a point, which would *appear* larger: the image of the pencil (with the lens in place) or the pencil (with the lens removed)? Explain how you can tell from the ray diagram.

2. If you were to measure the length of the pencil and the length of the image using a ruler, which would actually be larger? Explain how you can tell from the ray diagram.

⇨ Check your results for section II with a tutorial instructor.

III. A magnifying glass

A. Obtain a convex lens. Use the lens as a *magnifying glass,* that is, to make an object such as a pencil appear larger. Start with the lens very close to the object.

Which is farther from you: the image or the object?

Where is the *object* relative to the lens and its focal points? (For example, is the object distance *greater than, less than,* or *equal to* the focal length of the lens?)

B. Draw a ray diagram that shows how to determine the location of the image that you observed above. Your diagram need not be drawn exactly to scale, but should correctly show the location of the object relative to the observer and to the lens and its focal points.

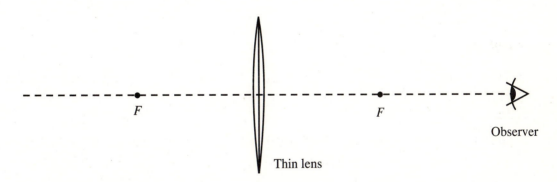

Thin lens

1. On the basis of your ray diagram, which is farther from the observer: the image or the object?

Is your answer consistent with your observations from part A? If not, resolve the inconsistency.

2. Does a magnifying glass simply make an object *appear* closer (*i.e.,* does it simply form an image of the object that is closer to you than the object itself)? If not, what does it do?

3. How can you tell from your ray diagram which would appear larger: the image of the pencil (with the lens in place) or the pencil (with the lens removed)?

IV. Magnification

A. The *lateral magnification*, m_1, is defined as $m_1 = h'/h$, where h' and h represent the heights of the image and object, respectively. By convention, h' and h have opposite signs when the image is inverted.

Does the value of the lateral magnification depend on the location of the observer? Explain.

Consider the two examples in this tutorial. In each case, is the *absolute value* of the lateral magnification *greater than, less than,* or *equal to* one? (If your answer depends on observer location, choose an observer who can see the entire image.)

Does the lateral magnification tell you whether the image will *appear* larger than the object without the lens? Explain why or why not.

B. The *angular magnification*, m_θ, is defined as $m_\theta = \theta'/\theta$, where θ' and θ represent the angular sizes of the image and object, respectively. By convention, θ' and θ have opposite signs when the image is inverted.

Does the value of the angular magnification depend on the location of the observer? Explain.

Consider the two examples in this tutorial. In each case, is the *absolute value* of the angular magnification *greater than, less than,* or *equal to* one? (If your answer depends on observer location, choose an observer who can see the entire image.)

Does the angular magnification tell you whether the image will *appear* larger than the object without the lens? Explain why or why not.

⇨ Check your answers for section IV with a tutorial instructor.

I. Periodic circular waves: single source

The circles at right represent wavefronts of a periodic circular wave in a portion of a ripple tank. The dark circles represent crests; the dashed circles, troughs. The diagram shows the locations of the wavefronts at one instant in time, as a photograph would.

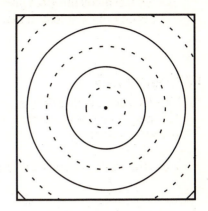

How, if at all, would the diagram differ:

• one-quarter period later? Explain.

• one period later? Explain.

II. Periodic circular waves: two sources

A. The diagram at right illustrates the wavefronts due to each of two small sources.

How do the frequencies of the two sources compare? Explain how you can tell from the diagram.

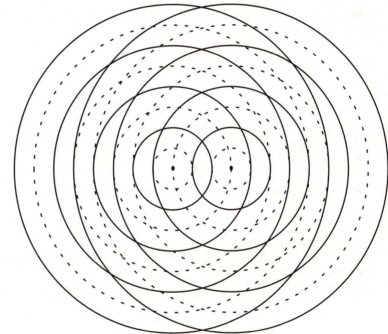

Are the two sources in phase or out of phase with respect to each other? Explain how you can tell from the diagram.

What is the source separation? Express your answer in terms of the wavelength.

Tutorials in Introductory Physics
McDermott, Shaffer, & P.E.G., U. Wash.

©Prentice Hall, Inc.
First Edition, 2002

B. Describe what happens at a point on the surface of the water where:

- a crest meets a crest

- a trough meets a trough

- a crest meets a trough

For each of the above cases, describe how your answer would differ if the amplitudes of the two waves were *not* equal. Explain your reasoning.

If the waves from two identical sources travel different distances to reach a particular point, the amplitudes of the waves from the two sources will not be the same at that point. For points that are sufficiently far from the sources, however, the difference in the amplitudes of the waves is small. For the remainder of this tutorial, we will *ignore* any such amplitude variations.

C. You have been provided a larger version of the diagram of the wavefronts due to two sources.

Use different symbols (or different colors) to mark the places at which *for the instant shown:*

- the displacement of the water surface is zero (*i.e.,* at its equilibrium level)
- the displacement of the water surface is the greatest above equilibrium
- the displacement of the water surface is the greatest below equilibrium

(*Hint:* Look for patterns that will help you identify these points.)

What patterns do you notice? Sketch the patterns on the diagram in part A.

D. The representation that we have been using indicates the shape of the water surface at one particular instant in time.

Consider a point on your diagram where a crest meets a crest.

How would the displacement of the water surface at this point change over time? (*e.g.,* What would the displacement be one-quarter period later? What would it be one-half period later?)

Consider what happens at a point on your diagram where a crest meets a trough.

How will the displacement of the water surface at this point change over time?

Tutorials in Introductory Physics
McDermott, Shaffer, & P.E.G., U. Wash.

©Prentice Hall, Inc.
First Edition, 2002

E. Suppose that a small piece of paper were floating on the surface of the water. Use your diagram to predict where the paper would move (1) the least and (2) the most.

F. Consider a point where the water surface remains undisturbed.

   1. Explain why that point *cannot* be the same distance from the two sources that we are considering.

      For the two sources that we are considering, by how much must the distances from that point to the two sources differ?

      Is there more than one possible value for the difference in distances? If so, list the other possible value(s) for the difference in distances. Explain.

   2. Choose a variety of points where the water surface remains undisturbed.

      For each of these points, determine the difference in distances from the point to the two sources. We will call this difference in distances ΔD. (This difference in distances is often called the *path length difference*.)

      Divide all of the points where the water surface remains undisturbed into groups that have the same value of ΔD. Label each group with the appropriate value of ΔD, in terms of the wavelength, λ.

      Justify the term *nodal lines* for groups of points that are far from the sources.

   3. Similarly, group the points where there is maximum constructive interference according to their value of ΔD. We will call these *lines of maximum constructive interference*.

      Label each group with the appropriate value of ΔD, in terms of the wavelength, λ.

   4. Label each of the nodal lines and lines of maximum constructive interference with the corresponding value of $\Delta\varphi$, the phase difference between the waves from the two sources.

⇨ Check your answers thus far with a tutorial instructor before continuing.

G. Imagine observing the waves from above the ripple tank. How, if at all, would the nodal lines and lines of maximum constructive interference change over time? Explain.

      What patterns and symmetries do you notice in the arrangement of the nodal lines and the lines of maximum constructive interference?

H. Each of the photographs at right shows a *part* of a ripple tank that contains two sources that are in phase.

For each of the photographs, identify:

• nodal lines
• the approximate locations of the sources
• the line that contains the two sources

Which of the two photographs more closely corresponds to the situation that you have been studying? Explain your reasoning.

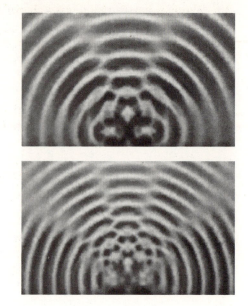

What difference(s) in the two situations could account for the difference in the number and the locations of the nodal lines?

I. Obtain a piece of paper and a transparency with concentric circles on them. The circles represent wavefronts generated by each of two point sources.

Suppose that the two sources are in phase and at the same location. Overlay the transparency on the paper to model this situation.

Explain why there are *no* nodal lines in this case.

Gradually increase the source separation until you first see nodal lines.

In the space at right, sketch the nodal lines and the lines of maximum constructive interference for this situation.

What is the source separation when this occurs?

Why can there be no nodal lines for a smaller source separation? Explain. *(Hint:* For a given source separation, what is the largest possible value of $\Delta D?$)

Continue to increase the source separation and investigate how the source separation affects the number of nodal lines and their locations.

⇨ Check your answers above with a tutorial instructor.

Tutorials in Introductory Physics   ©Prentice Hall, Inc.
McDermott, Shaffer, & P.E.G., U. Wash.   First Edition, 2002

I. Water waves incident on a single slit

A. Obtain a pan of water and form a barrier in it that has a wide slit as shown. Place a dowel in the water and gently rock it back and forth to generate straight wavefronts at a constant rate.

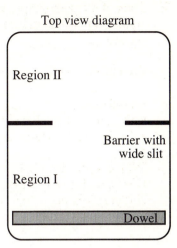

Top view diagram

Gradually decrease the width of the slit until it is completely closed. Observe the wavefronts in Region II as you make this change.

1. Describe how the shape of the wavefronts in Region II is affected as the width of the slit is decreased.

2. Compare the spacing of the wavefronts in the two regions (I and II). Is the spacing of the wavefronts in Region II affected by changing the width of the slit?

   Explain how your observation of the spacing of the wavefronts is consistent with the relative wave speeds in the two regions.

3. How, if at all, does the amplitude of the wave in Region II change when the slit is made slightly narrower? In particular, consider two cases in which:

   • the slit is initially very wide and is made slightly narrower.

   • the slit is initially very narrow and is made even narrower.

B. It is difficult to make periodic waves using the equipment at your table. Ask a tutorial instructor for photographs of periodic waves incident on slits of various widths.

1. Are the wavefronts in the photographs consistent with your observations above?

2. Identify the picture(s) in which the slit acts most like a point source of water waves. Explain.

   How could you modify the situation in this photograph in order to make the slit act more like a point source of waves?

3. Identify the photograph(s) in which the slit does not significantly affect the shape of the wavefronts.

   How could you modify the situation in this photograph so that the slit affects the wavefronts that pass through the slit to an even lesser extent?

Tutorials in Introductory Physics
McDermott, Shaffer, & P.E.G., U. Wash.

©Prentice Hall, Inc.
First Edition, 2002

As you have observed, the behavior of waves passing through a slit can depend on the size of the slit. For the remainder of this tutorial, we will consider the case of waves passing through two very narrow slits.

II. Water waves incident on two very narrow slits

For this part of the tutorial, you will not be asked to perform any experiments.

A. A periodic water wave is incident on a barrier with two identical narrow slits. Each slit is narrow enough so that it may be treated as a (single) source of circular wavefronts.

   Describe the shape of the wavefronts that emanate from each slit.

Top view diagram (not to scale)

B. Obtain an enlargement of the diagram at right that shows the wavefronts for the case in which the distance between the centers of the slits is 3λ.

   For this situation, which values of ΔD (the difference in distances from a point to each of the slits) correspond to (1) nodal lines and (2) lines of maximum constructive interference? Explain.

   At how many points along the line $X–X'$ in the diagram above is there (1) complete destructive interference and (2) maximum constructive interference? Mark the *approximate* locations of all of these points on the diagram above, and label each point with the corresponding value of ΔD. Assume that the tank is very wide and that the line $X–X'$ is very far from the slits.

C. Suppose that the width of *one* of the slits were decreased (without changing the distance between the centers of the slits). How, if at all, would this modification affect how much the water surface would move at the points you marked above? Explain your reasoning. (*Hint:* How can you use your observations from part A of section I in this case?)

   Thus far we have observed the behavior of *water waves* when they pass through narrow slits.

Below we investigate the behavior of *light* passing through two very narrow slits.

Tutorials in Introductory Physics
McDermott, Shaffer, & P.E.G., U. Wash.

©Prentice Hall, Inc.
First Edition, 2002

III. Light incident on two narrow slits

A. Red light from a distant point source is incident on a mask with two identical, narrow vertical slits. The photograph at right illustrates the pattern that appears at the center of a distant screen.

How does this pattern *differ* from what you would have predicted *if* you had used the idea that light travels in straight lines through slits?

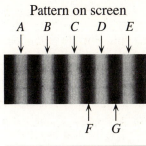

Pattern on screen

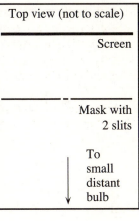

Top view (not to scale)

B. Compare the situation in part II (in which a periodic water wave was incident on two identical, narrow slits) to the experiment described above.

Which points along line *X–X'* in the ripple tank best correspond to:

• points of minimum intensity *(e.g., points F and G)?* Explain.

• points of maximum intensity *(e.g., points A–E)?* Explain.

For a point of minimum intensity *(e.g., points F and G)*, identify the quantity or quantities that are adding to zero at that point. Explain your reasoning.

C. For each of the lettered points, determine ΔD (in terms of λ) and $\Delta \varphi$, the phase difference between the waves. Record your answers below. *Note:* Point C is at the center of the screen.

	point A	point B	point C	point D	point E	point F	point G
ΔD							
$\Delta \varphi$							

D. Suppose that one of the slits were covered.

At which, if any, of the points A–G would the brightness increase? Explain.

At which, if any, of the points A–G would the brightness decrease? Explain.

In the space at right, sketch the pattern that would appear on the portion of screen shown in the above photograph when one of the slits is covered. Explain.

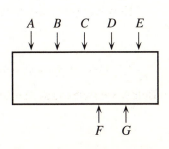

E. The pattern produced by red light passing through two very narrow slits has been reproduced at right.

Pattern on screen

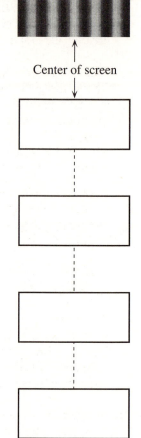

Center of screen

In each part below, suppose that a *single* change were made to the original apparatus. For each case, determine how, if at all, that change would affect the pattern on the screen. Sketch your predictions in the spaces provided.

1. the distance between the slits is decreased (without changing the width of the slits)

2. the screen is moved closer to the mask containing the slits

3. the wavelength of the incident light is decreased

4. the width of each slit is decreased (without changing the distance between the slits)

F. Consider point *B*, the first maximum to the left of the center of the screen.

Pattern on screen

B   C

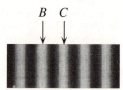

Suppose that the two slits are separated by 0.2 mm, that the screen is 1.2 m away from the slits, and that the distance from the center of the pattern (point *C*) to point *B* is 3.6 mm.

Use this information to determine the wavelength of the light. Describe any approximations that you make in answering this question.

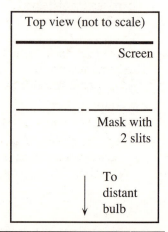

Top view (not to scale)

Screen

Mask with 2 slits

To distant bulb

Tutorials in Introductory Physics
McDermott, Shaffer, & P.E.G., U. Wash.

©Prentice Hall, Inc.
First Edition, 2002

I. Double-slit interference

A. Red light from a distant point source is incident on two very narrow identical slits, S_1 and S_2, separated by a distance d. The photograph at right illustrates the pattern that appears on a distant screen.

Double-slit pattern on screen

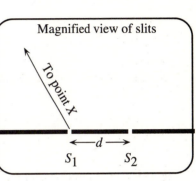

The magnified view shows the path from slit S_1 to point X, a point on the screen.

On the magnified view:

- Draw an arrow to show the direction from slit S_2 to point X.

- Identify and label the line segment of length ΔD that represents how much farther light travels from one slit than from the other to reach point X.

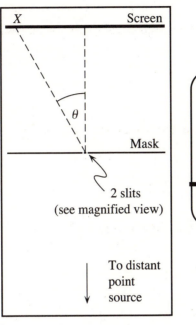

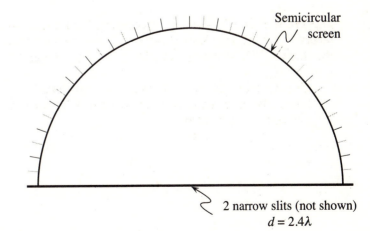

Magnified view of slits

B. In a previous homework, you found an expression for ΔD in terms of d and θ that was valid for points far from two point sources. Using that expression, write equations (in terms of λ, θ, and d) that you can use to calculate the angle(s) for which there will be:

- maximum constructive interference (*i.e.*, a *maximum*)

- complete destructive interference (*i.e.*, a *minimum*)

C. Suppose that the screen were semicircular, as shown.

On the diagram, mark the locations of *all* minima and maxima for the specific case $d = 2.4\lambda$.

Label each maximum and minimum with the corresponding value of:

- ΔD,
- θ, and
- $\Delta\varphi$, the phase difference between the waves from the two slits.

Semicircular screen

2 narrow slits (not shown)
$d = 2.4\lambda$

Tutorials in Introductory Physics
McDermott, Shaffer, & P.E.G., U. Wash.

©Prentice Hall, Inc.
First Edition, 2002

II. Three-slit interference

A. Consider a point *M* on the distant screen where there is a *maximum* due to the light from S_1 and S_2.

If a third slit were added as shown at right, would there *still* be maximum constructive interference at point *M?* Explain.

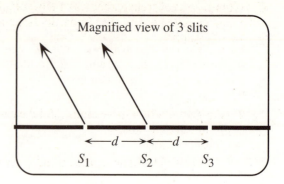

Magnified view of 3 slits

S_1 S_2 S_3

Suppose that more identical slits were added with all adjacent slits a distance *d* apart.

Would there still be maximum constructive interference at point *M?* Explain.

Let ΔD_{adj} represent the difference in distances from two adjacent slits to a location on the screen. For two slits, you found that any point of maximum constructive interference is farther from one of the slits by a whole number of wavelengths (*i.e.*, ΔD is 0, λ, 2λ,...).

For three or more evenly-spaced slits, what is the corresponding condition for locations of maximum constructive interference? Express this condition in terms of ΔD_{adj}.

We will call a location at which light from *all* of the slits is in phase a *principal maximum.*

B. Consider a point *N* on the screen where there is a *minimum* due to the light from S_1 and S_2.

Will the screen remain completely dark at point *N* after the third slit is added as shown above? If not, will point *N* be as bright as a principal maximum? Explain.

Tutorials in Introductory Physics
McDermott, Shaffer, & P.E.G., U. Wash.

©Prentice Hall, Inc.
First Edition, 2002

C. Obtain a set of transparencies of sinusoidal curves. Each transparency can represent the light from a single narrow slit. In particular, what quantity or quantities can these curves be used to represent?

Find a way to align three sinusoidal curves so they would add in a way that results in a minimum for three slits.

What is the smallest value of ΔD_{adj} that corresponds to a minimum for three slits?

Would twice this value also correspond to a minimum? three times? four times?

Write out the first few values of ΔD_{adj} that correspond to minima for three slits. Write out enough values to clearly indicate the pattern.

How many minima are there between adjacent principal maxima for three slits?

D. On the diagram at right, mark the locations of *all* minima and principal maxima for the specific case of three identical slits separated by a distance $d = 2.4\lambda$.

Label each minimum and principal maximum with the corresponding value of:

- ΔD_{adj},
- θ, and
- $\Delta \varphi_{adj}$, the phase difference between the waves from *adjacent* slits.

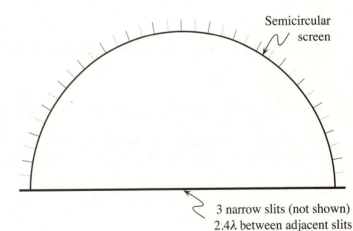

Semicircular screen

3 narrow slits (not shown)
2.4λ between adjacent slits

Compare and contrast this sketch with your sketch from part C of section I for the case of light incident on two slits separated by $d = 2.4\lambda$. Discuss the similarities and differences.

Tutorials in Introductory Physics
McDermott, Shaffer, & P.E.G., U. Wash.

©Prentice Hall, Inc.
First Edition, 2002

III. Multiple-slit interference

A. Suppose that coherent red light were incident on a mask with four narrow slits a distance d apart.

Use the transparencies of sinusoidal curves to find the smallest value of ΔD_{adj} that would correspond to a minimum for this case.

Which integer multiples of this value of ΔD_{adj} would correspond to other minima? Which would not?

Which values of ΔD_{adj} would correspond to the principal maxima?

How many minima would there be between adjacent principal maxima?

B. Generalize your results from the 2-slit, 3-slit, and 4-slit cases to determine the smallest value of ΔD_{adj} that would correspond to a minimum for the case of N identical, evenly-spaced slits.

Which integer multiples of this value of ΔD_{adj} would correspond to other minima? Which would not?

How many minima would there be between adjacent principal maxima?

C. Coherent red light is incident on a mask with two very narrow slits a distance d apart. The photograph at right illustrates the pattern that appears on a distant screen.

On the photograph, label each of the maxima and minima with the corresponding value of ΔD_{adj}.

Suppose that a *third* slit were added to the mask so that adjacent slits were separated by the same distance d as before.

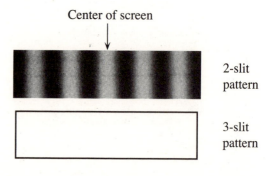

Center of screen

2-slit pattern

3-slit pattern

In the space provided above, sketch the pattern that you would expect to see on the same part of the screen. On your sketch, clearly label each minimum and principal maximum with the corresponding value of ΔD_{adj}.

Ask a tutorial instructor for photographs that illustrate the patterns that appear on a distant screen when light is incident on two masks: one with two slits and one with three slits.

Tutorials in Introductory Physics
McDermott, Shaffer, & P.E.G., U. Wash.

©Prentice Hall, Inc.
First Edition, 2002

D. How would the pattern differ if the mask contained *four* slits separated by the same distance *d* as before? Sketch your prediction in the space provided at right. Explain your reasoning.

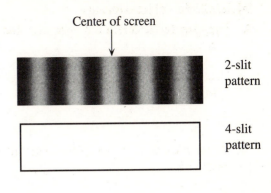

Center of screen

2-slit pattern

4-slit pattern

How would the pattern differ if the mask contained *five* slits separated by the same distance *d* as before? Explain your reasoning.

Ask a tutorial instructor for photographs that illustrate the patterns that appear on a distant screen when light is incident on masks with different numbers of slits.

Tutorials in Introductory Physics
McDermott, Shaffer, & P.E.G., U. Wash.

©Prentice Hall, Inc.
First Edition, 2002

I. Determining the location of the first minimum for many slits

A. Red light from a distant point source is incident on a mask with ten identical, evenly-spaced, very narrow slits. (See diagrams at right and below.)

On the magnified view below, label the line segment of length ΔD_{adj} that represents how much farther light must travel from slit 1 than from slit 2 to reach point X on a distant screen.

What is the *smallest* value of ΔD_{adj} that corresponds to a minimum for 10 slits? (Transparencies of sine curves are available in case you would like to review these concepts.)

The minimum that corresponds to this smallest value of ΔD_{adj} is called the *first minimum*.

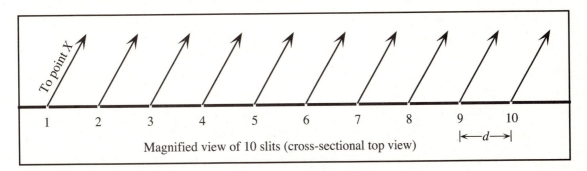

Magnified view of 10 slits (cross-sectional top view)

B. Suppose that point X marks the location of the first minimum on the screen.

How much farther (in terms of λ) does the light from slit 1 travel than the light from slit 3 in reaching point X? Explain.

C. Suppose that only slit 1 is uncovered, and all other slits 2–10 are covered. Which other slit could be uncovered so that the screen would be completely dark at point X? Explain.

Suppose that this *pair* of slits is uncovered, so that point X is completely dark. If slit 2 were now uncovered, would point X remain completely dark? If not, which other slit could also be uncovered (to pair with slit 2) so that point X once again becomes completely dark? Explain.

Tutorials in Introductory Physics
McDermott, Shaffer, & P.E.G., U. Wash.

©Prentice Hall, Inc.
First Edition, 2002

D. Show how you could group all ten slits into five pairs of slits so that the light waves from each pair add to zero at point X.

E. Suppose that the number of slits is doubled and the distance between adjacent slits is halved. (See below.) The new slits are labeled *1a–10a*. (The diagram uses the same scale as the preceding one.)

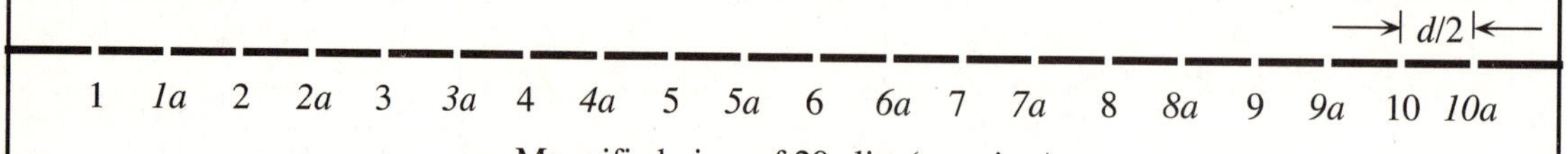

| 1 | *1a* | 2 | *2a* | 3 | *3a* | 4 | *4a* | 5 | *5a* | 6 | *6a* | 7 | *7a* | 8 | *8a* | 9 | *9a* | 10 | *10a* |

$\rightarrow$ $d/2$ $\leftarrow$

Magnified view of 20 slits (top view)

Would the first minimum in this case be located at the same angle θ as in part B? Explain.

F. If we continued to add slits in this way (*i.e.,* doubling the number of slits, but halving the distance between adjacent slits), would the angle to the first minimum change? Explain.

When the number of slits becomes very large as shown below, how can the slits be paired to determine the angle to the first minimum?

Magnified view of many, many slits (top view)

Tutorials in Introductory Physics
McDermott, Shaffer, & P.E.G., U. Wash.

©Prentice Hall, Inc.
First Edition, 2002

II. Motivation for a model for single-slit diffraction

The photograph below illustrates the pattern that appears on a distant screen when light from a distant point source passes through a single narrow vertical slit. This pattern is an example of a *single-slit diffraction pattern*.

A. How is this pattern *different* from what you would predict using the ideas developed in geometrical optics (*e.g.*, light travels in straight lines through slits)?

The presence of minima in a diffraction pattern suggests that diffraction is an interference phenomenon. We can model single-slit diffraction as follows: Consider the slit as consisting of many identical, very narrow, evenly-spaced "slits" that are so close to one another that the edges of these "slits" meet. The interference pattern produced by the light passing through the many "slits" approximates the single-slit diffraction pattern.

B. Consider the following dialogue:

Student 1: *"I don't see why there are minima when there's only a single slit—I think you need two waves to have destructive interference."*

Student 2: *"You can model the single slit as many identical smaller interfering 'slits,' each small enough to act like a point source. The first minimum occurs where the path length difference from the two 'slits' at the edges of the single slit is λ/2."*

Do you agree with student 2's response to student 1? Discuss your reasoning with your partners.

III. Applications of the model

A. The photograph at right shows the diffraction pattern produced on a distant screen by green light incident on a narrow slit.

Use an "✕" to mark the locations that correspond to the first minima.

Which locations correspond to higher-order minima?

Suppose that red light, instead of green light, were incident on the same slit.

Determine whether the angle to the first minimum for red light would be *greater than*, *less than*, or *equal to* that for green light. In the space below, draw diagrams that support your prediction, and explain your reasoning.

Green light incident on a narrow slit

To first minimum for green light

(Angle to first minimum exaggerated)

Red light incident on a narrow slit

Would you expect the locations of the higher-order minima to change? If so, how?

In the space below the photograph at the top of the page, show how the diffraction pattern would be different if red light, rather than green light, were incident on the narrow slit.

Obtain a color photograph that shows the diffraction patterns produced by red light and by green light on a narrow slit so that you may check your predictions.

⇨ Discuss your answers with a tutorial instructor.

Tutorials in Introductory Physics
McDermott, Shaffer, & P.E.G., U. Wash.

©Prentice Hall, Inc.
First Edition, 2002

B. The photograph at right shows
the diffraction pattern
produced by laser light
incident on a narrow slit.

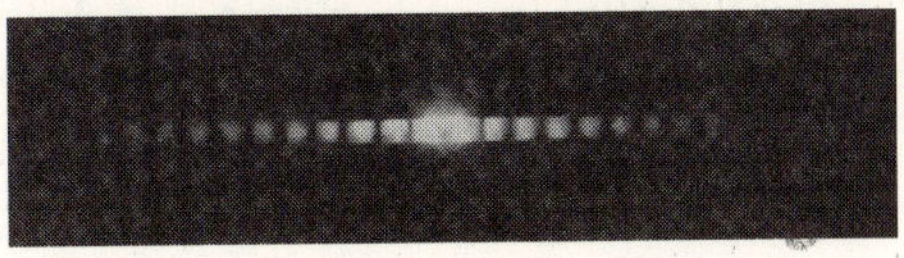

Narrow slit

Use the model that we have
developed to predict how the
pattern would change if the slit
were made *even narrower*.
Explain your reasoning and
sketch your prediction in the
space provided at right.

Even narrower slit

Ask a tutorial instructor for the photograph showing diffraction patterns produced by light
incident on a narrow slit and on an even narrower slit so that you may check your predictions.

C. Describe what you would see on the screen if the width of the slit were gradually decreased to
zero. Discuss your predictions with your partners.

D. If a diffraction pattern has several minima (like the patterns shown in this tutorial), is the
width of the slit *greater than*, *less than*, or *equal to* λ? Explain your reasoning.

E. In part A, you drew a diagram that showed how to find the angle to the first minimum for green light incident on a narrow slit. Use your diagram to determine whether the width of the slit was *greater than, less than,* or *equal to* the wavelength of the incident light in that case.

Is this comparison consistent with your answer to part D? If not, resolve the inconsistency.

F. Use the model that we have developed to write an equation that can be used to determine the angle to the first minimum in the case of single-slit diffraction with a slit of width *a*.

Explain how you can account for the fact that the above equation, which describes the location of a minimum in the case of single-slit diffraction, is similar in appearance to the equation that describes the location of a maximum in the case of two-source interference.

Tutorials in Introductory Physics
McDermott, Shaffer, & P.E.G., U. Wash.

©Prentice Hall, Inc.
First Edition, 2002

I. Single-slit diffraction

Monochromatic light from a distant point
source is incident on a mask that contains a
single narrow vertical slit. The photograph at
right shows the pattern produced on a distant
semicircular screen. The corresponding graph
of relative intensity is shown above the
photograph, where *relative intensity* is
intensity divided by the intensity at $\theta = 0$
(*i.e.*, $I(\theta)/I_{max}$).

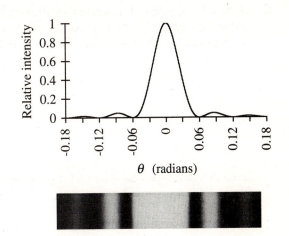

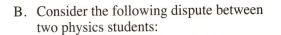

Pattern on screen due to *single* slit

A. The minima that occur in the case of a
single slit are called *diffraction minima*.
On the photograph and on the graph,
identify the locations of the diffraction
minima.

B. Consider the following dispute between
two physics students:

Student 1: *"In lab, I determined that the width of one of the slits that we used to study
single-slit diffraction was about 0.1 mm—that's definitely larger than λ."*

Student 2: *"You must have made a mistake. A diffraction pattern has minima only
when the slit width is less than λ."*

Do you agree or disagree with each of these students? Explain your reasoning.

II. Combined interference and diffraction

A. A second slit, identical in size to the first, is cut in the mask. The distance between the
centers of the slits, *d*, is equal to 50λ.

What would you see on the screen if the original slit were covered and the second slit were
uncovered?

B. Both slits are now uncovered.

For what angles will the light from each point on one slit be 180° out of phase with the light
from the corresponding point on the other slit? (*Hint:* For small angles, $\sin \theta \approx \theta$, where θ is
in radians.)

On the relative intensity graph above, clearly label these angles.

Tutorials in Introductory Physics
McDermott, Shaffer, & P.E.G., U. Wash.

©Prentice Hall, Inc.
First Edition, 2002

When the second slit is uncovered, will the intensity at the locations of the diffraction minima *increase, decrease,* or *stay the same?* Explain.

When the second slit is uncovered, how will the pattern on the screen change? In the space below, show how the pattern would be different.

	Pattern on screen due to *single* slit
	Pattern on screen due to *two* slits

⇨ Check your answers to part B with a tutorial instructor.

C. Suppose that the width of both slits, *a,* were gradually decreased (while keeping the distance between the centers of the slits the same).

Which minima would move as *a* is decreased?

Choose two or more relative intensity graphs below that illustrate such a change. (Enlargements of these graphs have also been provided.)

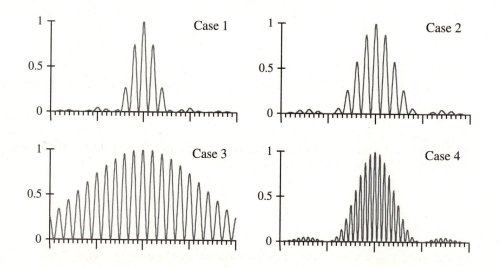

D. Suppose instead that the distance between the centers of the slits, *d,* were gradually decreased (while keeping the widths of the slits the same).

Which minima would move as *d* is decreased?

Choose two or more relative intensity graphs in part C above that illustrate such a change.

The minima that occur when only one slit is open are called diffraction minima. The minima that occur where the light from each point on one slit is 180° out of phase with the light from the corresponding point on the other slit are called *interference minima*.

E. The four graphs from part C that show relative intensity versus angle for two slits are given below. In each case:

   • Clearly label (1) the interference minima that are closest to the center of the pattern and (2) the diffraction minima that are closest to the center of the pattern.

   • Sketch the graph of relative intensity that would result if one of the slits were covered.

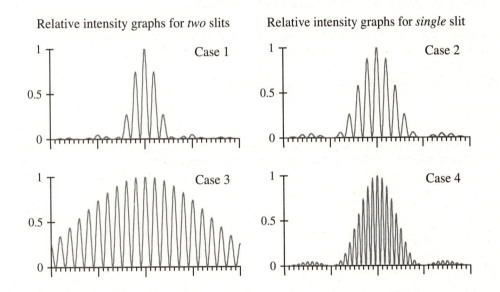

For each of the four cases, is your relative intensity graph consistent with the minima that you identified? If not, resolve any inconsistencies.

⇨ Check your answers to parts C–E with a tutorial instructor.

©Prentice Hall, Inc.
First Edition, 2002

F. Consider the relative intensity graph shown at right.

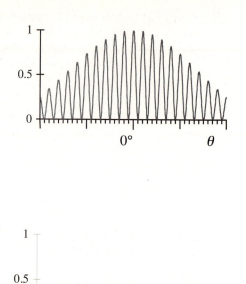

Suppose that both slits were made narrower (without changing the distance between the centers of the slits).

On the graph at right, indicate how the relative intensity would change. Explain.

Suppose that after gradually narrowing both slits, one of the slits were then covered. On the axes provided, sketch the relative intensity graph for this case.

How does your graph compare to what you would expect for a *point source?* If it is different, how could you modify the physical situation so that the relative intensity graph better approximates that due to a point source?

In order for the relative intensity graph to be a good approximation of that due to a point source, how must the width of the slit compare to λ? Explain your reasoning.

III. Quantitative predictions

Consider the following relative intensity graph for a double-slit experiment. The wavelength of the light used was $\lambda = 633$ nm.

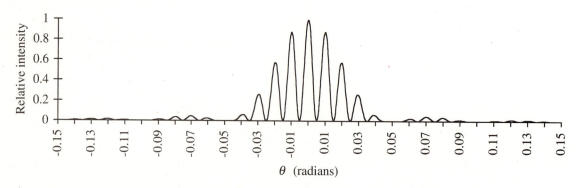

A. Determine the width of the slits and the distance between the slits. Clearly indicate which features of the graph you are using.

Compare your results with those obtained by your partners. If your answers are different, resolve any discrepancies.

Tutorials in Introductory Physics
McDermott, Shaffer, & P.E.G., U. Wash.

©Prentice Hall, Inc.
First Edition, 2002

B. Consider the following comment made by a student:

> *"To determine slit width, I used the first minimum, at θ = 0.005 radians, and to determine the distance between the slits, I used the first maximum, at θ = 0.01 radians."*

What is the flaw in the reasoning used by this student? Explain your reasoning.

C. You may have already noticed that the maxima are (approximately) 0.01 radians apart, except that there are no maxima at $\theta = 0.05$ radians or $\theta = 0.10$ radians.

How can you account for these "missing" maxima? (*Hint:* Consider how the relative intensity graph would be different if the width of the slits were decreased.)

Are your answers from part A consistent with your answer above? If not, resolve the inconsistency.

I. Transmission and reflection at a boundary

The sketches below show a pulse approaching a boundary between two springs. In one case, the pulse approaches the boundary from the left; in the other, from the right. The springs are the same in both cases, and the linear mass density is greater for the spring on the right than for the spring on the left.

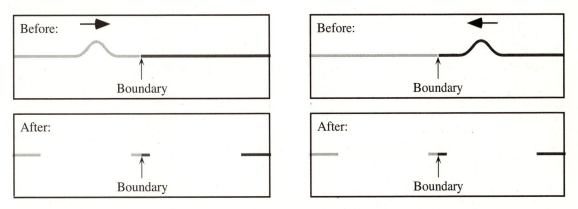

Complete the sketches to show the shape of the springs a short time after the trailing edge of the pulse shown has reached the boundary. Be sure to show correctly (1) the relative widths of the pulses and (2) which side of the spring each pulse is on. (Ignore relative amplitudes.)

Compare your diagrams with those of your partners. Resolve any inconsistencies.

II. Thin-film interference

You may have observed that when a beam of light strikes a piece of glass, it is partially reflected and partially transmitted, similar to the behavior of a pulse on a spring when it reaches the junction between two connected springs.

In this tutorial, we consider a beam of light in air ($n = 1$) incident on a soap film ($n = {}^4/_3$). We make an analogy between this situation and a pulse incident on a boundary between two springs of different mass densities.

A. In this analogy, would the soap film better correspond to the spring with the larger linear mass density or the smaller linear mass density? Explain your reasoning.

Discuss your reasoning with your partners.

B. When comparing two materials of different indices of refraction, the material with the higher index of refraction is sometimes said to be more "optically dense" than the other. Is this terminology consistent with the analogy that you made in part A?

Tutorials in Introductory Physics
McDermott, Shaffer, & P.E.G., U. Wash.

©Prentice Hall, Inc.
First Edition, 2002

C. Consider light incident on a *thin* soap film, as illustrated in the cross-sectional side view diagram at right.

The soap film is supported by a loop (not shown), which is held vertically. Only a small portion of the film has been shown; the thickness of the film is greatly exaggerated.

In answering the following questions, use an analogy between this situation and the connected springs in parts A and B.

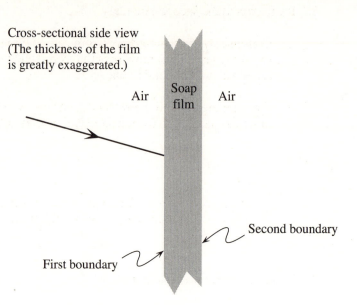

Cross-sectional side view
(The thickness of the film is greatly exaggerated.)

Air   Soap film   Air

Second boundary

First boundary

1. Reflection and transmission at the first boundary

a. On the diagram, draw rays that correspond to the light that is transmitted and reflected at the first boundary (on the left).

b. Is the frequency of the transmitted wave (in the film) *greater than, less than,* or *equal to* the frequency of the incident wave (in the air)?

c. Is the wavelength of the transmitted wave (in the film) *greater than, less than,* or *equal to* the wavelength of the incident wave (in the air)?

d. For light incident on the first boundary, would the reflection at this boundary be more like reflection from a *fixed end* or from a *free end?* Explain.

e. On the basis of your answers above:

At the first boundary, would the reflected wave be *in phase* or *180° out of phase* with the incident wave (*i.e.,* is there a phase change upon reflection)?

At the first boundary, would the transmitted wave be *in phase* or *180° out of phase* with the incident wave (*i.e.,* is there a phase change upon transmission)?

Tutorials in Introductory Physics
McDermott, Shaffer, & P.E.G., U. Wash.

©Prentice Hall, Inc.
First Edition, 2002

2. Reflection at the second boundary

   a. Continue the transmitted ray (from part 1) through the film to the second boundary (on the right). Then draw rays that correspond to the light that is transmitted and reflected at the second boundary.

   b. For light incident on the second boundary, would the reflection at this boundary be more like reflection from a *fixed end* or from a *free end*? Explain.

   At the second boundary, would the reflected wave be *in phase* or *180° out of phase* with the incident wave (in the film)?

3. Transmission at the first boundary

   Continue the reflected ray from part 2 through the film back to the first boundary. Then draw rays that correspond to the light that is transmitted and reflected at this boundary.

   Would there be a phase change on transmission at this boundary?

D. Light of frequency $f = 7.5 \times 10^{14}$ Hz is incident on the left side of the soap film.

   Determine the numerical values of the:

   • frequency of the wave in soap film (in Hz)

   • wavelength in air (in nm)

   • wavelength in film (in nm)

E. Suppose that an observer were located on the left side of the soap film in part C.

   Which of the rays that you drew could reach this observer?

   How would these rays be different if the light were incident at essentially normal incidence?

III. A film of non-uniform thickness

A soap film supported by a vertical loop has settled and is thinner at the top than at the bottom. Light of frequency $f = 7.5 \times 10^{14}$ Hz is incident on the film ($n = {}^4\!/\!_3$) at essentially normal incidence.

Cross-sectional side view (not to scale)

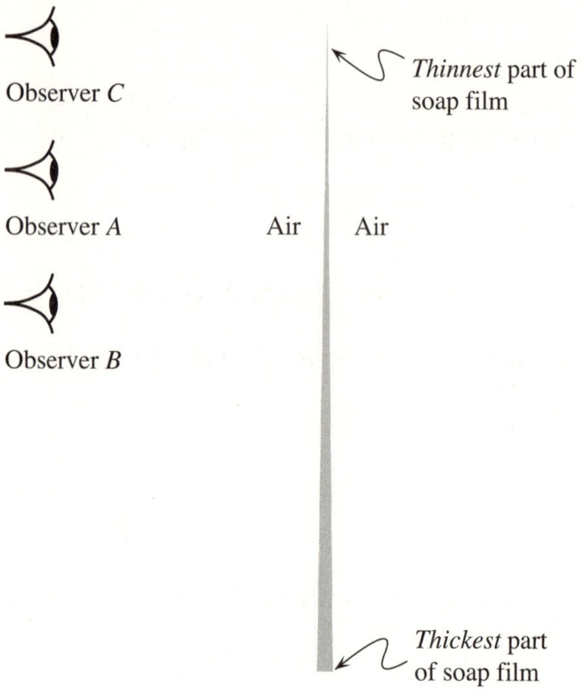

Observer C

Observer A

Observer B

Air   Air

Thinnest part of soap film

Thickest part of soap film

A. Observer A is looking at the part of the film that is 75 nm thick.

Consider two reflected rays that reach observer A, similar to the rays that you identified in part E of section II.

1. How much farther does one of these rays travel than the other in reaching observer A in *mm?*

2. What is the phase difference between these rays? (Be sure to take into account the phase changes that you identified in part C of section II as well as any phase difference due to path length difference.)

3. Is observer A looking at a region of *maximum brightness, minimum brightness,* or *neither?* Explain your reasoning.

B. Observer B is looking at the part of the film that is 150 nm thick.

Is this observer looking at a region of *maximum brightness, minimum brightness,* or *neither?* Explain your reasoning.

C. Observer *C* is looking at the thinnest part of the film, where the film is *extremely* thin. To this observer, would the film appear *bright* or *dark?* Explain your reasoning.

D. Describe the appearance of the film as a whole.

⇨ Check your answers to parts A–D with a tutorial instructor.

E. What are the three smallest film thicknesses for which there would be maximum *constructive* interference?

What are the three smallest film thicknesses for which there would be maximum *destructive* interference?

F. The thickness of the film is 1650 nm at the bottom of the film, where the film is the thickest.

1. Would this part of the film appear *bright, dark,* or *in between?* Explain.

2. Suppose that the frequency of the incident light were increased.

How, if at all, would the appearance of the thinnest part of the film change?

Would the number of bright and dark regions *increase, decrease,* or *stay the same?* Explain your reasoning.

I. Polarization of light

A. Look at the room lights through one of the polarizing filters provided.

Describe how the filter affects what you see. Does rotating the filter have an effect?

B. Hold a second polarizing filter in front of the first, and look at the room lights again.

Describe how the filter affects the light that you see. How does rotating one of the filters with respect to the other affect what you see?

On the basis of your observations, why is it appropriate to use the term *filter* to describe these pieces of apparatus?

How is the behavior of the polarizing filters *different* from the behavior of colored acetate filters?

You have learned that light may be thought of as a wave consisting of oscillating electric and magnetic fields. If the electric field in all parts of a light beam oscillates along a single axis, the light beam is said to be *linearly polarized,* or simply, *polarized*. For example, the diagram at right represents a polarized light wave moving in the *x*-direction in which the electric field oscillates only along the *y*-axis. By convention, the direction along which the electric field oscillates (in this case, the *y*-direction) is called the *direction of polarization* of a light beam. If the electric field oscillates in different, random directions within the same light beam, that beam is said to be *unpolarized*.

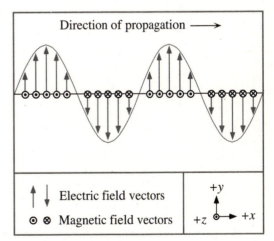

Tutorials in Introductory Physics
McDermott, Shaffer, & P.E.G., U. Wash.

©Prentice Hall, Inc.
First Edition, 2002

II. Polarizing filters

The light transmitted by a polarizing filter (or *polarizer*) depends upon the relative orientation of the polarizer and the electric field in the light wave. Every polarizer has a *direction of polarization*, which is often marked by a line drawn on it. The electric field of the transmitted wave is equal to the component of the electric field of the incident wave that is *parallel* to the direction of polarization of the polarizer.

A. Do the room lights produce polarized light? Explain how you can tell from your observations.

B. Suppose that you had two marked polarizers (*i.e.*, their directions of polarization are marked).

Predict how you should orient the polarizers with respect to one another so that the light transmitted through the polarizers would have (1) *maximum* intensity or (2) *minimum* intensity. Discuss your reasoning with your partners and then check your predictions.

When two polarizers are oriented with respect to each other such that the light transmitted through them has minimum intensity, the polarizers are said to be *crossed*.

C. Suppose that you had a polarizer with its direction of polarization marked. How could you use this polarizer to determine the direction of polarization of another (unmarked) polarizer? Explain your reasoning.

D. A beam of light is incident on a polarizer, as shown in the side view diagram below. The direction of polarization of the light makes an angle θ with respect to the polarizer's direction of polarization. (See front view diagram.) The amplitude of the electric field of the incident light is E_o. The magnetic field (not shown) has an amplitude B_o.

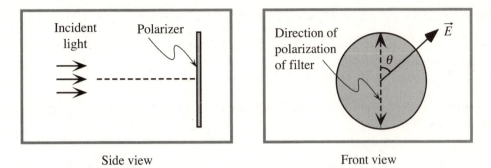

Side view   Front view

The vector $\vec{E}$ represents the electric field of the incident light at the front surface of the polarizer at a particular time. Resolve $\vec{E}$ into two components: one that is *transmitted* by the polarizer and one that is *absorbed* by the polarizer.

What is the *direction* of the electric field of the transmitted light? How, if at all, is it different from the direction of the electric field of the incident light? Explain.

Write an expression for the *amplitude* of the electric field of the transmitted light, in terms of E_o and θ.

Write an expression for the *amplitude* of the magnetic field of the transmitted light, in terms of B_o and θ. Explain your reasoning.

Write an expression for the *intensity* of the transmitted light in terms of I_o, the intensity of the incident light, and θ. Show all work. (*Hint:* If the amplitude of the electric field were reduced by a factor of two, by what factor would the intensity be reduced?)

⇨   Check your results from part D with a tutorial instructor.

E. An observer is looking at a light source through two polarizers as shown in the side view diagram at right. The polarizers are crossed, that is, they are oriented so that the light transmitted through them has *minimum* intensity.

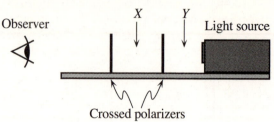

1. Suppose that a third polarizer were inserted at the position marked *X*, shown above.

   Predict how, if at all, this change would affect the intensity of the light reaching the observer. Does your answer depend on the orientation of the third polarizer? Discuss your reasoning with your partners.

   Check your prediction experimentally. (Ask a tutorial instructor to show you the equipment that you need in order to do so.) If your prediction was incorrect, identify those parts of your prediction that were wrong.

   How can you apply your results from part D to help you account for your observations? Support your answer with one or more diagrams.

2. Suppose that instead a third polarizer were inserted at the position marked *Y*, shown above.

   Predict how, if at all, this change would affect the intensity of the light reaching the observer. Does your answer depend on the orientation of the third polarizer? Discuss your reasoning with your partners.

F. Consider a beam of *unpolarized* light that is incident on a polarizer. What is the intensity of the transmitted light in terms of I_o, the intensity of the incident light? (*Hint:* We can think of unpolarized light as equal amounts of light that are polarized *parallel* and *perpendicular* to the direction of polarization of the polarizer.)

Tutorials in Introductory Physics
McDermott, Shaffer, & P.E.G., U. Wash.

©Prentice Hall, Inc.
First Edition, 2002

Selected topics

I. Applying Newton's laws to liquids

A rectangular container filled with water is at rest on a table as shown. Two imaginary boundaries that divide the water into three layers of equal volume have been drawn in the diagram. (No material barrier separates the layers.)

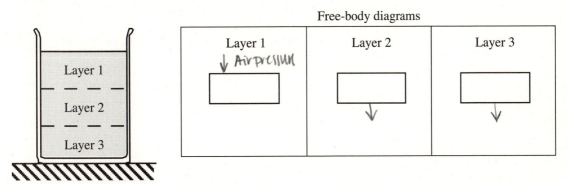

Free-body diagrams

A. For each layer, draw a free-body diagram in the space provided. Be sure to indicate on your diagram the surface on which each *contact* force is applied. (This is usually done by placing the tip of the arrow that represents the force at that surface.)

The label for each force should indicate:
- the type of force,
- the object on which the force is exerted, and
- the object exerting the force.

B. Rank the magnitudes of all the vertical forces you have drawn in the three diagrams from largest to smallest. Explain how you determined your ranking.

$$L_3 , L_2 , L_1$$

How does the weight of layer 1 compare to that of layer 3?

$$L_1 < L_3$$

A liquid in which equal volumes have equal weight regardless of depth (*i.e.*, the density does not vary) is referred to as *incompressible*. Assume that all liquids in this tutorial are incompressible.

C. Imagine that a small hole is opened in the container wall near the bottom of each layer.

1. Predict what will happen to the water near each hole. Explain.

   The water on top shoot less distance than the water on bottom

2. Check your prediction by observing the demonstration. Record your observations. (A sketch may be helpful.)

   What do your observations suggest about: (1) the *existence* of horizontal forces on the three layers of water in part A? (2) the *relative magnitudes* of the horizontal forces on the three layers?

   If necessary, revise your free-body diagrams in part A so that they are consistent with your answers.

Tutorials in Introductory Physics
McDermott, Shaffer, & P.E.G., U. Wash.

©Prentice Hall, Inc.
First Edition, 2002

II. Pressure and force

A. Recall the relationship between force and pressure. (Consult your textbook if necessary.) Below we will apply this relationship to the three layers from part I.

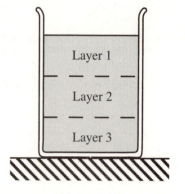

1. Which *force* would you use to determine the *pressure* at the bottom of layer 2? (There may be more than one correct answer.) Explain your reasoning. (*Hint:* Refer to your free-body diagrams from section I. Which forces are exerted at the bottom of layer 2?)

    atmospheric pressure + acceleration due to gravity

2. Which *area* would you use to determine the pressure at the bottom of layer 2? Explain.

3. Suppose that you wanted to determine the pressure at a point in the center of layer 2. For what object(s) would you draw a free-body diagram? Which force and which area would be useful in determining the pressure?

    layer 1 + layer 2

B. Suppose you wanted to determine the pressure at the top surface of layer 1. Which force would you use to determine this pressure? If necessary, modify your free-body diagrams to include this force. Be sure to label your diagram to indicate the object that exerts this force.

    atmospheric pressure + gravity

Three points, *L, M,* and *N,* are marked at the bottom of the three layers.

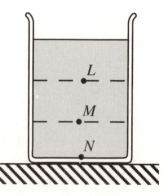

C. Rank the pressures at points *L, M,* and *N.* Explain how your answer is consistent with your ranking of forces in section I.

    N > M > L

The pressure *P* at a point in an incompressible liquid is often described mathematically as $P = P_o + \rho g h$.

D. Is your ranking in part C consistent with this equation? (*Hint:* At what point is $h = 0$? What is the pressure at that point?)

    yes, atm. pressure

III. Pressure as a function of depth

The container at right is filled with water and is at rest on a table. An imaginary boundary that outlines a small volume of water has been drawn in the diagram. Treat this small volume of water as a single object.

A. Draw a free-body diagram for the small volume of water in the space below the figure.

B. Compare the magnitudes of the *horizontal* forces that you have drawn.

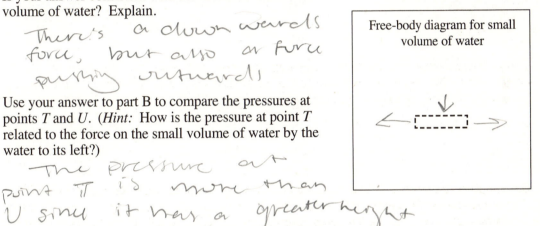

Is your answer consistent with the motion of the small volume of water? Explain.

There's a downwards force, but also a force pushing outwards

Free-body diagram for small volume of water

C. Use your answer to part B to compare the pressures at points T and U. (*Hint:* How is the pressure at point T related to the force on the small volume of water by the water to its left?)

The pressure at point T is more than U since it has a greater height

D. Rank the pressures at points Q, R, S, T, and U. Explain.

Q <

E. Consider the following student dialogue:

Student 1: *"The pressure at a point is equal to the weight of the water above divided by the area. Therefore the pressure at point R is greater than the pressure at point S because there's no water above point S."*

Student 2: *"I agree. The pressure is $P_0 + \rho g h$, and h is zero for point S and greater than zero for point R. Therefore, the pressure at R must be greater."*

Do you agree with either student? Explain your reasoning.

⇨ Discuss your reasoning with a tutorial instructor.

IV. Pressure in a U-tube

A U-shaped tube is filled with water as shown.

A. Rank the pressures at points A through F. Explain.

$$F < A = E < B < C < D$$

Is your ranking consistent with the equation $P = P_0 + \rho g h$? Explain.

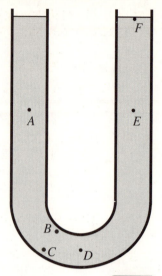

B. The right end of the tube is now sealed with a stopper. The water levels on both sides remain the same. There is no air between the stopper and the water surface.

1. Does the pressure at points A and D *increase, decrease,* or *remain the same?* Explain.

2. Is the pressure at point E *greater than, less than,* or *equal to* the pressure at point $D?$

   Does the difference in pressure ΔP_{DE} between points D and E change when the stopper is added? Explain.

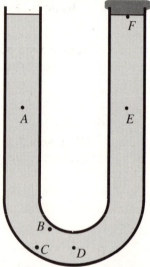

3. Is the pressure at point F *greater than, less than,* or *equal to* atmospheric pressure?

   Is the force exerted by the rubber stopper on the water surface on the right *greater than, less than,* or *equal to* the force exerted by the atmosphere on the water surface on the left?

C. A syringe is used to remove some water from the left side of the U-tube. The water level on the left side is seen to be lowered, but the water level on the right does not change.

Consider the following student dialogue:

Student 1: *"The pressure at point F must now be higher than atmospheric pressure because the water there is being pushed up against the stopper."*

Student 2: *"I think that the pressure at point E must be the same as at point A because they are at the same level. These points are both at atmospheric pressure. So the pressure at point F is lower than atmospheric pressure because we know that pressure gets less as you go up."*

Student 3: *"But water is more dense than air so the pressure at F cannot be less than atmospheric pressure."*

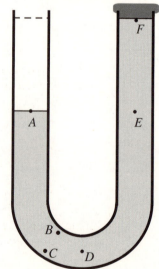

With which student(s), if any, do you agree?

Tutorials in Introductory Physics
McDermott, Shaffer, & P.E.G., U. Wash.

©Prentice Hall, Inc.
First Edition, 2002

I. Buoyant force

A. A cubical block is observed to float in a beaker of water. The block is then held near the center of the beaker as shown and released.

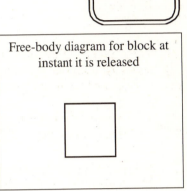

1. Describe the motion of the block after it is released.

2. In the space provided, draw a free-body diagram for the block at the instant that it is released. Show the forces that the water exerts on each of the surfaces of the block separately.

    Make sure the label for each force indicates:

    • the type of force,
    • the object on which the force is exerted, and
    • the object exerting the force.

    Free-body diagram for block at instant it is released

3. Rank the magnitudes of the vertical forces in your free-body diagram. If you cannot completely rank the forces, explain why you cannot.

    Did you use the relationship between pressure and depth to compare the magnitudes of any of the vertical forces? If so, how?

    Did you use information about the motion of the block to compare the magnitudes of any of the vertical forces? If so, how?

4. In the box at right, draw an arrow to represent the vector sum of the forces exerted on the block by the surrounding water. How did you determine the direction?

    Sum of forces on block by water

    Is this vector sum the *net force* on the block? (Recall that the net force is defined as the vector sum of *all* forces acting on an object.)

    Is the magnitude of the sum of the forces exerted on the block by the water *greater than, less than,* or *equal to* the weight of the block? Explain.

Tutorials in Introductory Physics
McDermott, Shaffer, & P.E.G., U. Wash.

©Prentice Hall, Inc.
First Edition, 2002

B. The experiment is repeated with a second block that has the same volume and shape as the original block. However, this block is observed to sink in water.

1. In the space provided, draw a free-body diagram for the block at the instant it is released. As before, draw the forces exerted on each surface of the block by the water.

2. Compare the free-body diagram for the block that sinks to the one you drew in part A for the block that floats.

Which forces are the same in magnitude and which are different? (*Hint:* How does the pressure at each surface of this block compare to the pressure at the corresponding surface of the block in part A?)

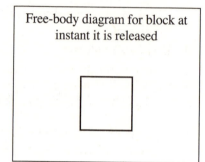

Free-body diagram for block at instant it is released

Do any forces appear on one diagram but not on the other?

3. In the space provided, draw an arrow to represent the vector sum of the forces exerted on the block by the water.

How does this vector compare to the one you drew for the block that floats? (Consider both magnitude and direction.)

Sum of forces
on block by water

C. Imagine that you were to release the block from part B at a much greater depth. State whether each of the following forces on the block would be *greater than, less than,* or *equal to* the corresponding force on the block in part B above:

1. the upward force on the bottom surface on the block.

2. the downward force on the top surface of the block.

3. the vector sum of the forces on the block by the surrounding water. (*Hint:* Does the difference between the pressures at the top and bottom surfaces of the block change?)

The vector sum of the forces exerted on an object by a surrounding liquid is called the *buoyant force*. This force is customarily represented by a single arrow on a free-body diagram.

Tutorials in Introductory Physics
McDermott, Shaffer, & P.E.G., U. Wash.

©Prentice Hall, Inc.
First Edition, 2002

D. In general, does the buoyant force on an object that is completely submerged in an incompressible liquid depend on:

- the *mass* or *weight* of the object?

- the *depth* below the surface at which the object is located?

- the *volume* of the object?

⇨ Check your answers with a tutorial instructor before continuing.

II. Displaced volume

Consider two blocks of the same size and shape: one made of aluminum; the other, of brass. Both blocks sink in water. The aluminum block is placed in a graduated cylinder containing water. The volume reading increases by 3 mL.

A. By how much does the volume reading increase when the brass block is placed in the cylinder? (Assume that no water leaves the cylinder.) Explain.

When an object is placed in a graduated cylinder of liquid, the increase in the volume reading is called the *volume of liquid displaced* by the object.

B. Does the volume of water displaced by a *completely submerged* object depend on

- the *mass* or *weight* of the object?

- the *depth* below the surface at which the object is located?

- the *volume* of the object?

- the *shape* of the object?

III. Archimedes' principle

According to *Archimedes' principle*, the magnitude of the buoyant force exerted on an object by a liquid is equal to the weight of the volume of that liquid displaced by the object.

A. Consider the following statement made by a student:

> *"Archimedes' principle simply means that the weight of the water displaced by an object is equal to the weight of the object itself."*

Do you agree with the student? Explain.

IV. Sinking and floating

A. A rectangular block, A, is released from rest at the center of a beaker of water. The block accelerates upward.

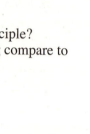

1. At the instant it is released, is the buoyant force on block A *greater than, less than,* or *equal to* its weight? Explain.

2. When block A reaches the surface, it is observed to float at rest as shown. In this final position, is the buoyant force on the block *greater than, less than,* or *equal to* the weight of the block? (*Hint:* What is the net force on the object?)

3. Are your answers to the questions above consistent with Archimedes' principle? (*Hint:* How does the volume of water displaced when the block is floating compare to that displaced when it was completely submerged?)

B. A second block, B, of the same size and shape as A but slightly greater mass is released from rest at the center of the beaker. The final position of this block is shown at right.

How does the buoyant force on block B compare to the buoyant force on block A:

• at the instant they are released? Explain.

• at their final positions? Explain.

C. A third block, C, of the same size and shape as A and B but with slightly greater mass than block B is released from rest at the center of the beaker. Two students predict the final position of the block and draw the sketch at right.

Student 1: *Since this block is heavier than block B, it will not go up as high after it is released, as shown at right.*

Student 2: *Yes, I agree, the buoyant force is slightly less than the weight of this block, so it should come to rest a bit below the surface.*

Explain what is *wrong* with each statement and with the diagram.

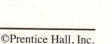

Student drawing

Tutorials in Introductory Physics
McDermott, Shaffer, & P.E.G., U. Wash.

©Prentice Hall, Inc.
First Edition, 2002

I. Pressure

A cylinder contains an ideal gas that is at room temperature. The cylinder is sealed with a piston of mass M and cross-sectional area A that is free to move up or down without friction. No gas can enter or leave the cylinder. The piston is at rest. Atmospheric pressure (*i.e.*, the pressure of the air surrounding the cylinder) is P_o.

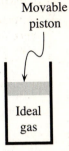

Movable piston

Ideal gas

A. In the space provided, draw a free-body diagram for the piston.

Make sure the label for each force indicates:

- the type of force,
- the object on which the force is exerted, and
- the object exerting the force.

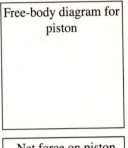

Free-body diagram for piston

B. In the space provided, draw an arrow to indicate the direction of the net force on the piston. If the net force is zero, state so explicitly.

C. Is the force exerted on the piston by the gas *inside* the cylinder *greater than*, *less than*, or *equal to* the force exerted on the piston by the air *outside* the cylinder? Explain.

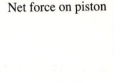

Net force on piston

Write an equation that relates all the forces on your free-body diagram. (*Hint:* How are these forces related to the net force?)

D. Is the pressure of the gas in the cylinder *greater than*, *less than*, or *equal to* atmospheric pressure? Explain.

Determine the value of the pressure of the gas in the cylinder in terms of the given quantities. (*Hint:* Which of the forces that act on the piston can you use to find the pressure of the gas?)

E. A second cylinder contains a different sample of ideal gas at room temperature, as shown at right. The two cylinders and their pistons are identical.

Is the pressure of the gas in the second cylinder *greater than*, *less than*, or *equal to* the pressure in the cylinder above? If you cannot tell, state so explicitly. Explain.

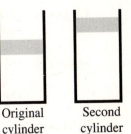

Original cylinder

Second cylinder

⇨ Check your answer with a tutorial instructor before you continue.

Tutorials in Introductory Physics
McDermott, Shaffer, & P.E.G., U. Wash.

©Prentice Hall, Inc.
First Edition, 2002

II. Pressure and temperature

A. A cylinder of the type described in section I contains a fixed amount of gas. Initially, it is in thermal equilibrium with an ice-water bath. The pressure, volume, and temperature of the gas are $P_{initial}$, $V_{initial}$, and $T_{initial}$, respectively.

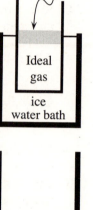

Movable piston

Ideal gas

ice water bath

The cylinder is then removed from the ice water and placed into boiling water. After the system has come to thermal equilibrium with the boiling water the pressure, volume, and temperature are P_{final}, V_{final}, and T_{final}.

1. Is T_{final} *greater than, less than,* or *equal to* $T_{initial}$?

2. Is P_{final} *greater than, less than,* or *equal to* $P_{initial}$? Explain.

   Is your answer consistent with your answer to part D of section I? If not, resolve any inconsistencies.

boiling water bath

3. Is V_{final} *greater than, less than,* or *equal to* $V_{initial}$? Explain.

   Is your answer consistent with the ideal gas law *(i.e., the relationship PV = nRT)?* If not, resolve any inconsistencies.

B. In the process you considered in part A above, which variables are held constant and which are allowed to change? Explain how you can tell.

C. Consider the following student dialogue.

   Student 1:   *"According to the ideal gas law, the pressure is proportional to the temperature. Since I increased the temperature of the gas, the pressure must go up."*

   Student 2:   *"That's right. Since no gas entered or left the system, the volume did not change. So the pressure must have increased."*

   Do you agree with either of the students? Explain your reasoning.

   ⇨   Check your reasoning with a tutorial instructor before you continue.

III. *PV* diagrams

Ideal gas processes are often represented graphically. For instance, a *PV diagram* is a graph of pressure *versus* volume for a given sample of gas. A single point on the graph represents simultaneously measured values of pressure and volume. These values define a *state* of the gas.

A. Sketch the process described in section II on the *PV* diagram provided. Label the initial and final states of the gas.

Is your sketch consistent with your answer in part B of section II? Explain.

B. The same sample of ideal gas is used for a new experiment. The pressure and volume of the gas are measured at several times. The values of *P* and *V* are recorded on the diagram shown at right.

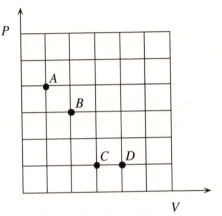

1. Rank the temperatures of the gas in states *A, B, C,* and *D* from largest to smallest. If any two temperatures are the same, state so explicitly.

2. Is your ranking consistent with the ideal gas law?

3. Is it possible for the gas to be in a state in which it has the same volume as in state *B* and the same temperature as in state *A*? If so, mark the location of the state on the *PV* diagram. If not, explain why not.

⇨ Check your reasoning with a tutorial instructor before you continue.

Tutorials in Introductory Physics
McDermott, Shaffer, & P.E.G., U. Wash.

©Prentice Hall, Inc.
First Edition, 2002

IV. Avogadro's number

A. Two identical cylinders of the type described above contain hydrogen and oxygen, respectively. Both cylinders have been in the same room for a long time. Their pistons are at the same height.

1. Compare the volumes of the gases in the two cylinders. Explain.

Hydrogen   Oxygen

2. Compare the temperatures of the gases in the two cylinders. Explain.

3. Compare the pressures in the two cylinders. Explain.

4. Compare the number of moles in the two cylinders. Explain.

Is your answer consistent with the ideal gas law?

B. A student looks up the molar masses and finds the values 2 g (for H_2) and 32 g (for O_2).

1. Give an *interpretation* of these two numbers. (*Note:* A formula is not considered an interpretation.)

2. Compare the masses of the gas samples in the two containers. Explain.

C. Consider the following student discussion.

Student 1:   *"Since hydrogen molecules are so much smaller than oxygen molecules, there should be more of them in the same volume."*

Student 2:   *"No, since n = 2 for hydrogen, and n = 32 for oxygen, there must be more oxygen molecules."*

Find the flaws in the statements of both students. Explain.

⇨ Check your reasoning with a tutorial instructor.

I. Work

A. Recall the definition of work done on an object by an agent that exerts a force on that object. (You may wish to consult your textbook.)

In the spaces provided, sketch arrows representing (1) a force exerted on an object and (2) the displacement of that object for cases in which the work done by the agent is:

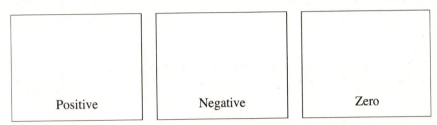

| Positive | Negative | Zero |

In each case, does your sketch represent the *only* possible relative directions of the force and displacement vectors? If so, explain. If not, sketch at least one other possible set of vectors.

B. A block is pushed by a hand as it moves from the bottom to the top of a frictionless incline. The block is speeding up at a constant rate.

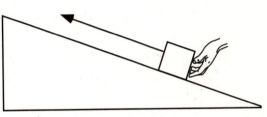

Frictionless incline

1. In the space provided, draw a free-body diagram for the block.

   Make sure the label for each force indicates:

   • the type of force,
   • the object on which the force is exerted, and
   • the object exerting the force.

2. In the space provided, draw an arrow to show the direction of the net force on the block.

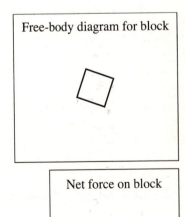

Free-body diagram for block

Net force on block

3. State whether the following quantities are *positive, negative,* or *zero*. In each case, explain your reasoning.

   • the work done on the block by the *hand*

   • the work done on the block by the *Earth*

   • the work done on the block by the *incline*

4. Is there work done on the hand by the block in this motion? If so, is this work *positive, negative,* or *zero?* Explain.

5. The *work-kinetic energy theorem* states that the change in kinetic energy of a rigid body is equal to the net work done on that body. Explain how your answers to part 3 are consistent with this theorem. (*Hint:* The *net work* is the sum of the works done by all forces exerted on an object.)

6. Which, if any of your answers in part 3 would be different if the block were being pushed up the incline with constant speed?

   Describe the net work done on the block in that case.

C. An ideal gas is contained in a cylinder that is fixed in place. The cylinder is closed by a piston as shown in the diagram at right. There is no friction between the piston and the cylinder walls.

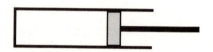

1. Describe the direction of the force that the piston exerts on the gas.

   Does your answer depend on whether the piston is moving?

2. How could the piston move so that the work it does on the gas is:

   • positive?

   • negative?

   Do your answers depend on your choice of coordinate system?

3. In each of the two cases in part 2, is there work done on the piston by the gas? If so, how is that work related to the work done on the gas by the piston? (Consider both sign and absolute value.)

   ⇨ Check your answers with a tutorial instructor before you continue.

II. Work and internal energy

A. Imagine that the cylinder from section I is thermally isolated from its surroundings by placing it in an insulating jacket. The piston is pressed inward to the position shown at right. We will refer to this compression as process 1.

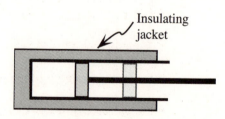

Insulating jacket

Is the work done on the gas by the piston *positive*, *negative*, or *zero*?

Tutorials in Introductory Physics
McDermott, Shaffer, & P.E.G., U. Wash.

©Prentice Hall, Inc.
First Edition, 2002

In thermal physics, we are often interested in the *internal energy* (E_{int}) of a system. The internal energy of an ideal gas is proportional to the temperature and the number of moles of the gas. The internal energy can change when energy is exchanged with the system's environment (*e.g.*, objects that are outside the system of interest). The case above is one in which the internal energy of a gas changes due to work done on the gas (the system) by the piston (an agent external to the system). When such a system is thermally isolated, the change in internal energy of the system is equal to the net work done on it:

$$\Delta E_{int} = W_{\text{on system}} \qquad \text{(for a thermally isolated system)}$$

B. 1. Does the internal energy of a gas in an insulated cylinder *increase, decrease,* or *remain the same* when the piston is pushed inward? Explain.

2. Does the temperature of the gas change? Explain.

C. Two students are discussing process 1:

Student 1: *"The volume of the gas decreases, but the pressure increases. Therefore, by the ideal gas law, the temperature must remain the same."*

Student 2: *"But I know the temperature goes up. The volume is less, and therefore the particles collide more often with one another."*

Neither student is correct. Find the flaws in the reasoning of each student. Explain.

⇨ Check your reasoning with a tutorial instructor before you continue.

III. Heat

A. Imagine that the cylinder from section II is no longer thermally insulated, and the piston is locked in place. The gas is initially at room temperature.

The cylinder is then placed into boiling water and reaches thermal equilibrium with the water. We refer to this process as process 2.

1. In process 2, do the following quantities *increase, decrease,* or *remain the same?* Explain.

- the temperature of the gas

- the internal energy of the gas

- the pressure of the gas

- the volume of the gas

Tutorials in Introductory Physics
McDermott, Shaffer, & P.E.G., U. Wash.

©Prentice Hall, Inc.
First Edition, 2002

2. Sketch process 2 on the *PV* diagram at right.

3. Is there any work done on the gas in process 2? Explain. Is
    your answer consistent with your *PV* diagram?

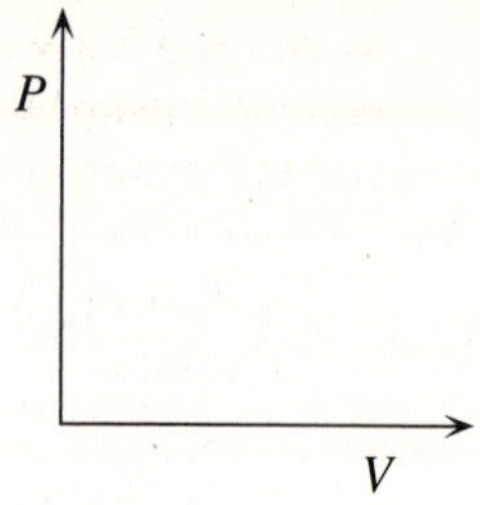

The energy transfer that takes place in this process is called *heat transfer*. In this process, if the heat transferred to the gas *(Q)* is greater than zero, the internal energy of the gas will increase.

B. In process 2, is the heat transfer to the gas *positive, negative, or zero?* Explain.

C. In process 2, is the heat transfer to the boiling water *positive, negative, or zero?* Explain.

IV. Heat, work, and internal energy

The *first law of thermodynamics* states that the change in internal energy of a closed system is equal to the sum of the net work done on the system and the heat transferred to the system:

$$\Delta E_{int} = Q + W_{\text{on system}}$$

A. Explain how you could write this law in terms of the work done *by* the system on its environment.

How does your textbook express the first law of thermodynamics?

B. In process 1 (section II) you did not need to consider heat transfer. What feature of the experiment prevented heat transfer to the gas?

C. In process 2 (section III) you did not need to consider work. What feature of the experiment prevented work from being done on the gas?

⇨ Check your reasoning with a tutorial instructor before you continue.

D. The cylinder, with the piston still locked in place, is now immersed in a mixture of ice and water and allowed to come to thermal equilibrium with the mixture. The piston is then moved inward very slowly, in such a way that the gas is always in thermal equilibrium with the ice-water mixture. We will refer to this slow compression of the gas as process 3.

  1. In process 3, do the following quantities *increase, decrease,* or *remain the same?* Explain.

   • the volume of the gas

   • the temperature of the gas

   • the internal energy of the gas

   • the pressure of the gas

  2. Sketch process 3 on the *PV* diagram provided.

  3. Determine whether the following quantities are *positive, negative,* or *zero:*

   • the work done on the gas in process 3 (Explain your reasoning by referring to a force and a displacement.)

   • the heat transfer to the gas in process 3

  4. Are your answers above consistent with the first law of thermodynamics? Explain.

E. How does the compression in process 3 differ from the compression in process 1? Explain.

F. A student is considering process 3:

   "The temperature doesn't change; it is an isothermal process. Therefore, the heat transfer must be zero."

   Do you agree with this student? Explain.

I. Review of two-slit interference of light

Light of wavelength λ from a distant point source is incident on two very narrow slits, S_1 and S_2, a distance d apart. (See diagrams at right.) The photograph above right shows the pattern seen on a distant screen.

Pattern on distant screen

Screen   point X

θ

Mask

2 slits
(see enlarged view)

Point source
(far from slits)

TOP VIEW *(not to scale)*

Magnified view of slits

to (distant) point X

θ

d

S_1 S_2

A. In the magnified view of the slits, an arrow is drawn showing the direction from slit S_1 to an arbitrary point on the screen, point X. On the magnified view:

  • draw an arrow to indicate the approximate direction from slit S_2 to the *distant* point X.

  • identify and label the line segment that represents the path length difference from the slits to point X.

For small angles θ (where θ is measured in radians), what is the approximate path length difference?

B. For what values of the path length difference (written in terms of λ) will there be:

  • maximum constructive interference *(i.e., a maximum)?*

  • complete destructive interference *(i.e., a minimum)?*

C. Suppose that a *single* change were made to the apparatus (keeping the distance between the mask and the screen fixed), resulting in the new pattern shown.

  1. Are the angles to the interference maxima in the new pattern *greater than, less than,* or *equal to* those in the original pattern? Explain how you can tell from the photographs.

Original pattern on screen

New pattern on screen

Center of screen

Tutorials in Introductory Physics
McDermott, Shaffer, & P.E.G., U. Wash.

©Prentice Hall, Inc.
First Edition, 2002

2. If the wavelength of light *(λ)* was the *only* quantity changed, determine (i) whether λ was *increased* or *decreased,* and (ii) whether it was changed by a factor that was *greater than, less than,* or *equal to* 2. Explain how you can use your results from parts A and B to justify your answer.

3. If the slit separation *(d)* was the *only* quantity changed, determine (i) whether *d* was *increased* or *decreased,* and (ii) whether it was changed by a factor that was *greater than, less than,* or *equal to* 2. Explain how you can use your results from parts A and B to justify your answer.

⇨   Check your reasoning with a tutorial instructor.

II. Two-slit interference of electrons

A beam of electrons is accelerated through a potential difference, *V.* The beam is incident on two narrow slits. The photograph shows the pattern seen on a phosphorescent screen placed far from the slits. (When an electron hits a small portion of the screen, that portion of the screen glows.)

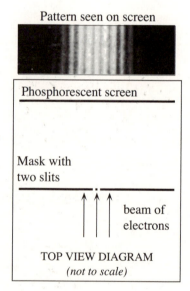

A. Which is a better model for how the electrons behave in this case: that they propagate in straight lines through the slits, or that they propagate like waves? Explain how you can tell.

B. Suppose that the above experiment were repeated but with the electrons accelerated through a potential difference of 0.5*V* instead of *V.*

1. Predict whether the bright regions on the screen would *move closer together, move farther apart,* or *stay at the same locations.* Discuss your reasoning with your partners.

Obtain a figure that shows how the interference pattern would change if the accelerating voltage were halved so that you may check your prediction.

2. On the basis of the figure, would you conclude that halving the accelerating voltage changes the wavelength of the electron wave?

    If so: Does the wavelength increase or decrease?
    Does the wavelength change by a factor that is *greater than, less than,* or *equal to* 2? Explain how you can tell from the figures.

    If not: Explain how you can tell that the wavelength did not change.

3. How would decreasing the accelerating voltage by a factor of one-half affect each of the quantities listed below? In particular, determine (i) whether each quantity would *increase* or *decrease,* and (ii) whether each quantity would change by a factor that is *greater than, less than,* or *equal to* 2. Explain your reasoning in each case.

    *   the kinetic energy of each electron that reaches the slits

    *   the momentum of each electron that reaches the slits

    *   the de Broglie wavelength of each electron that reaches the slits

4. Are your answers to parts 2 and 3 regarding the de Broglie wavelength of the electron consistent? If not, resolve any inconsistencies.

    Now that you have worked through parts 2 and 3, review your answer to part 1. Do you still agree with your earlier reasoning? If not, how would you revise it?

©Prentice Hall, Inc.
First Edition, 2002

C. Suppose you were to perform the electron interference experiment described in part A.

Describe two independent methods that you could use in order to determine the de Broglie wavelength of the electrons. Include in your descriptions the measurements you would need to make and the steps you would need to follow in each case.

III. Application: Davisson-Germer experiment

Monoenergetic electrons are incident on a nickel crystal. It is observed that intense scattering occurs at angles θ according to the Bragg condition, $2d \sin\theta = n\lambda$. (See diagram at right.)

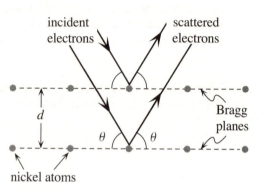

A. Use trigonometry to show that the path length difference between the two scattered beams shown is equal to $2d \sin\theta$. Show all work.

B. Suppose that this experiment were repeated, each time with a *single* change made to the apparatus. For each change below, determine whether each of the angles θ at which intense scattering occurs would become *larger, smaller,* or *stay the same*. Explain your reasoning in each case.

1. The kinetic energy of the incident electrons is decreased.

2. The electrons are replaced with neutrons, with each neutron having the same speed as each of the original electrons.

Tutorials in Introductory Physics
McDermott, Shaffer, & P.E.G., U. Wash.

©Prentice Hall, Inc.
First Edition, 2002

I. *I-V graphs for the photoelectric effect experiment*

In the experiment shown at right, an ammeter is connected in series with an evacuated tube containing two electrodes (A and B). The combination is placed in parallel with a voltmeter and a variable resistor. A source of monochromatic light is directed toward electrode B.

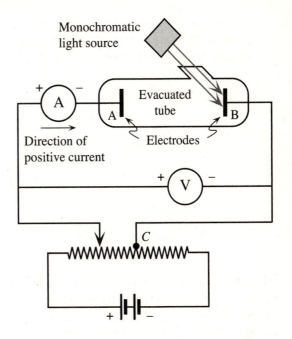

A. How does the voltmeter reading compare to the potential difference across the electrodes? Explain.

If the sliding lead from electrode A were connected at point *C* along the resistor, would the voltmeter reading be *positive, negative,* or *zero?* Explain.

(*Hint:* Imagine disconnecting the ammeter and evacuated tube from the rest of the circuit, and answering the same question.)

How would you adjust the sliding connection from electrode A in order to make the potential difference across the electrodes ($\Delta V_{BA} = V_A - V_B$) become (i) more and more positive? (ii) more and more negative? Explain.

B. The electrodes are made of aluminum, with a work function $\Phi = 4.2$ eV. The monochromatic light source emits light with a frequency of 1.5×10^{15} Hz. (Recall that $h = 4.14 \times 10^{-15}$ eV-s.)

In this case, would electrons be ejected from electrode B? If so, what would be the maximum kinetic energy of the electrons that are ejected? If not, explain why not.

1. Suppose that the sliding lead from electrode A is connected at point *C*.

Would the ammeter reading be *positive, negative,* or *zero?* Explain your reasoning.

Tutorials in Introductory Physics
McDermott, Shaffer, & P.E.G., U. Wash.

©Prentice Hall, Inc.
First Edition, 2002

2. Suppose that the potential difference across the electrodes (ΔV_{BA}) is gradually *increased* from zero to +8.0 V.

In this case, would the electrons be attracted *toward electrode A, attracted toward electrode B,* or *neither?* Explain. (*Hint:* If the potential difference is positive, what is the direction of the electric field between the electrodes?)

How would the ammeter reading change as the potential difference increases? Explain. (*Hint:* Were *all* the ejected electrons reaching electrode A in the situation in part 1?)

Monochromatic light source

A

Evacuated tube

A

B

Direction of positive current

Electrodes

+ V −

C

+ −

3. Suppose instead that the potential difference (ΔV_{BA}) is gradually *decreased* from zero to −8.0 V.

How would the current through the ammeter change in this case? Explain.

Would the current ever reach zero? If so, at what value of the potential difference would the current become zero? If not, explain why not.

Would the current ever become *negative?* Explain why or why not.

C. In the space at right, draw a graph of *current through the ammeter* versus *potential difference across the electrodes.* Assume that the light source and the electrodes are the same as in part B.

Is your *I-V* graph consistent with your answers in part B? If not, resolve the inconsistencies.

I (arbitrary units)

V (volts)

⇨ Check your *I-V* graph with a tutorial instructor before starting section II.

Tutorials in Introductory Physics
McDermott, Shaffer, & P.E.G., U. Wash.

©Prentice Hall, Inc.
First Edition, 2002

II. Predicting how changes in intensity, frequency, and work function affect *I-V* graphs

Obtain a handout that shows a typical *I-V* graph for the photoelectric effect experiment. Carefully copy the graph in the spaces provided on this page and the next. Be sure to show that the *stopping voltage* is equal to – 2.0 V. (The *stopping voltage* is determined by setting the voltage across the electrodes to zero and then making the voltage more and more negative. The first negative value for which the current becomes zero is called the *stopping voltage*.)

A. Suppose that the intensity of the light were increased (while the wavelength of the light remained the same).

1. In the space at right, *predict* the resulting *I-V* graph. (If possible, use different color inks for the original and modified graphs.)

   Explain the reasoning you used in drawing the new graph.

2. To help you check your graph, consider how the change described above would affect:

   • the maximum value of the current through the ammeter. Explain.

   • the value of the potential difference at which the current becomes non-zero. Explain.

3. Is your graph in part 1 consistent with your answers in part 2? If not, resolve any inconsistencies.

⇨ Check your *I-V* graph above with a tutorial instructor before you continue.

4. Consider the statement below:

   "In the original situation there was no current when V = -2 V. If the intensity of the light source is increased the total energy of the photons increases. This means the ejected electrons have more energy, so a voltage of -2 V isn't enough to 'stop' the current."

   Do you agree or disagree with this statement? Explain your reasoning.

Tutorials in Introductory Physics
McDermott, Shaffer, & P.E.G., U. Wash.

©Prentice Hall, Inc.
First Edition, 2002

B. Suppose that the frequency of the light were increased. (Assume that the intensity of the light is also adjusted so that the maximum current remains at the same value as in the original graph.)

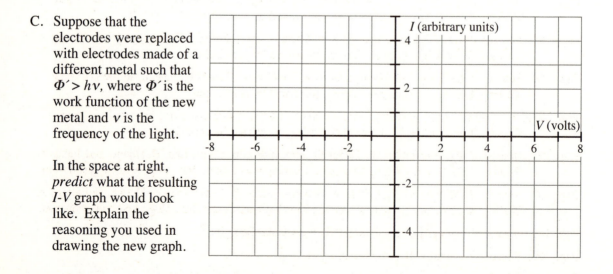

1. In the space at right, *predict* the resulting *I-V* graph. Explain the reasoning you used in drawing the new graph.

2. To help you check your graph, consider how the change described above would affect:

    • the energy of each photon incident on the electrode B. Explain.

    • the value of the potential difference at which the current becomes non-zero. Explain.

3. Is your graph in part 1 consistent with your answers in part 2? If not, resolve any inconsistencies.

    ⇨ Check your *I-V* graph above with a tutorial instructor.

C. Suppose that the electrodes were replaced with electrodes made of a different metal such that $\Phi' > h\nu$, where Φ' is the work function of the new metal and ν is the frequency of the light.

    In the space at right, *predict* what the resulting *I-V* graph would look like. Explain the reasoning you used in drawing the new graph.

    ⇨ Check your *I-V* graph with a tutorial instructor.

Credits: